The Oil Industry and Windfall Profits

THE OIL INDUSTRY AND WINDFALL PROFITS

MARIA B. COSWELL
EDITOR

Nova Science Publishers, Inc.
New York

For permission to use material from this book please contact us:
Telephone 631-231-7269; Fax 631-231-8175
Web Site: http://www.novapublishers.com

LIBRARY OF CONGRESS CATALOGING-IN-PUBLICATION DATA
Available upon request
ISBN 978-1-60692-840-0

Published by Nova Science Publishers, Inc. ✝ New York

CONTENTS

PREFACE

Over the past 10 years, surging crude oil and petroleum product prices have increased oil and gas industry revenues and generated record profits, particularly for the top five major integrated companies (also known as the "super-majors"): Exxon-Mobil, Royal Dutch Shell, BP, Chevron, and Conoco/Phillips. These companies, which reported a predominant share of those profits, generated more than $100 billion in profits on nearly $1.5 trillion of revenues in 2007. From 2003 to 2007, revenues increased by 51%; net income (profits) increased by 85%. Being largely price-driven, with no increase in output, and with little new production resulting from increased oil industry investment, many believe that a portion of the increased oil industry income over this period represents a windfall and unearned gain, i.e., income not earned by any additional effort on the part of the firms, but due primarily to record crude oil prices, which are set in the world oil marketplace. Numerous bills have been introduced in the Congress over this period to impose a windfall profits tax (WPT) on oil. This new book discusses these options.

Chapter 1 – In April 1980, the federal government enacted the crude oil windfall profit tax on the U.S. oil industry. The main purpose of the tax was to recoup for the federal government much of the revenue that would have otherwise gone to the oil industry as a result of the decontrol of oil prices. Supporters of the tax viewed this revenue as an unearned and unanticipated windfall caused by high oil prices, which were determined by the OPEC (Organization of Petroleum Exporting Countries) cartel.

Despite its name, the windfall profit tax (WPT) was actually an excise tax, not a profits tax, imposed on the difference between the market price of oil and an adjusted base price. While most domestically produced oil was subject to the tax (about 2/3 in 1985), the remaining 1/3 that was tax-exempt was significant (1.3

billion barrels in 1985, or 360,000 barrels per day). The $80 billion in gross revenues generated by the WPT between 1980 and 1988 was significantly less than the $393 billion projected. Due to the deductibility of the WPT against the income tax, cumulative net WPT revenues were about $38 billion, significantly less than the $175 billion projected. This report presents estimates of the amount of foregone oil production from 1980-1986 due to the WPT under three alternative supply price responses, reflecting three different assumptions about the price elasticity of the domestic oil supply function, a critical factor (statistic) in estimating lost oil output and increased import dependence. From 1980 to 1988, the WPT may have reduced domestic oil production anywhere from 1.2% to 8.0% (320 to 1,269 million barrels). Dependence on imported oil grew from between 3% and 13%. The tax was repealed in 1988 because (1) it was an administrative burden to the Internal Revenue Service (IRS), (2) it was a compliance burden to the oil industry, (3) due to low oil prices, the tax was generating little or no revenues in 1987 and 1988, and (4) it made the United States more dependent on foreign oil. The depressed state of the U.S. oil industry after 1986 also contributed to the repeal decision.

Reinstating the windfall profit tax would reduce recent oil industry windfalls due to high crude and petroleum prices but could have several adverse economic effects. If imposed as an excise tax, the WPT would increase marginal production costs and be expected to reduce domestic oil production and increase the level of oil imports, which today is at nearly 60% of demand. Crude prices would not tend to increase. Some have proposed an excise tax on both domestically produced and imported oil as a way of mitigating the negative effects on petroleum import dependence. Such a broad-based WPT would tend to reduce import dependence, but it would lead to higher crude oil prices and likely to oil industry profits, potentially undermining its original goals. Because the pure corporate profits tax is relatively neutral in the short run — few, if any, price and output effects occur because marginal production costs are unchanged in the short run — a possible option would be a corporate income surtax on the upstream operations of crude oil producers. Such a tax that would recoup any recent windfalls with less adverse economic effects; imports would not increase because domestic production would remain unchanged. In the long run, such a tax is a tax on capital; it reduces the rate of return, thus reducing the supply of capital to the oil industry.

Chapter 2 – Over the past 10 years, surging crude oil and petroleum product prices have increased oil and gas industry revenues and generated record profits, particularly for the top five major integrated companies (also known as the "super-majors"): Exxon- Mobil, Royal Dutch Shell, BP, Chevron, and Conoco/Phillips. These companies, which reported a predominant share of those profits, generated more than $100 billion in profits on nearly $1.5 trillion of revenues in 2007. From 2003 to 2007, revenues increased by 51%; net income (profits) increased by 85%. Oil output by the five majors over this time period declined by more than 2%, from 9.85 to 9.63 million barrels per day. Being largely price-driven, with no increase in output, and with little new production resulting from increased oil industry investment, many believe that a portion of the increased oil industry income over this period represents a windfall and unearned gain, i.e., income not earned by any additional effort on the part of the firms, but due primarily to record crude oil prices, which are set in the world oil marketplace.

Numerous bills have been introduced in the Congress over this period to impose a windfall profits tax (WPT) on oil. Most of the bills were introduced in the 109th and 1 10th Congresses, after the enactment of the Energy Policy Act of 2005, which provided oil and gas industry tax incentives, in addition to the industry's traditional tax subsidies. An excise-tax based WPT would tax only domestic production, and like the one in effect from 1980-1988, would increase marginal oil production costs, which theoretically could reduce domestic oil supply, and raise petroleum imports, making the United States more dependent on foreign oil, undermining goals of energy independence and energy security. By contrast, an income-tax based WPT would be more economically neutral (less distortionary) in the short-run: sizeable revenues could be raised without reducing domestic oil supplies. Neither the excise- tax based or income-tax based WPT are expected to have significant price effects: neither tax would increase the price of crude oil, which means that refined petroleum product prices, such as pump prices, would likely not tend to increase.

In lieu of these two types of WPT, an administratively simple way of increasing the tax burden on the oil industry, and therefore recouping some of any excess or windfall profits, particularly from major integrated producers, would raise the corporate tax rate by, for instance, repealing or reducing the domestic manufacturing activities deduction under IRC § 199. This deduction is presently 6% of a firm's net income) and is available generally to all domestic manufacturing businesses (service firms are excluded), including almost all oil firms. Repealing this deduction for the major integrated oil companies, and freezing it at 6% for the remaining qualifying oil companies is estimated by the Joint Committee on Taxation to generate about $10 billion over 10 years.

Chapter 3 – The CLEAN Energy Act of 2007 (H.R. 6) was introduced by the House Democratic leadership to revise certain tax and royalty policies for oil and natural gas and to use the resulting revenue to support a reserve for energy efficiency and renewable energy. Title I proposes to repeal certain oil and natural gas tax subsidies, and use the resulting revenue stream to support the reserve. The Congressional Budget Office (CBO) estimates that Title I would repeal about $7.7 billion in oil and gas tax subsidies over the 10-year period from 2008 through 2017. In House floor debate, opponents argued that the cut in oil and natural gas subsidies would dampen production, cause job losses, and lead to higher prices for gasoline and other fuels. Proponents counterargued that record profits show that the oil and natural gas subsidies were not needed. The bill passed the House on January 18 by a vote of 264-123. This report presents a detailed review of oil and gas tax subsidies, including those targeted for repeal by H.R. 6.

The Energy Policy Act of 2005 (EPACT05, P.L. 109-58) included several oil and gas tax incentives, providing about $2.6 billion of tax cuts for the oil and gas industry. In addition, EPACT05 provided for $2.9 billion of tax increases on the oil and gas industry, for a net tax increase on the industry of nearly $300 million over 11 years. Energy tax increases comprise the oil spill liability tax and the Leaking Underground Storage Tank financing rate, both of which are imposed on oil refineries. If these taxes are subtracted from the tax subsidies, the oil and gas refinery and distribution sector received a net tax increase of $1,356 million ($2,857 million minus $1,501 million).

EPACT05 was approved and signed into law at a time of very high petroleum and natural gas prices and record oil industry profits. The House approved the conference report on July 28, 2005, and the Senate on July 29, 2005, clearing it for the President's signature on August 8 (P.L. 109-58). However, the tax sections originated in the106th Congress, with its effort in 1999 to help the ailing domestic oil and gas producing industry, particularly small producers, deal with depressed oil prices. Subsequent price spikes prompted concern about insufficient domestic energy production capacity and supply. All the early bills appeared to be weighted more toward stimulating the supply of conventional fuels, including capital investment incentives to stimulate production and transportation of oil and gas.

In addition to the tax subsidies enacted under EPACT05, the U.S. oil and gas industry qualifies for several other targeted tax subsidies (FY2006 revenue loss estimates appear in parenthesis): (1) percentage depletion allowance ($1 billion); (2) expensing of intangible drilling costs for successful wells and non-geological and geophysical costs for dry holes, including the exemption from the passive loss limitation rules that apply to all other industries ($1.1 billion); (3) a tax credit for small refiners of low-sulfur diesel fuel that complies with Environmental Protection Agency (EPA) sulfur regulations ($ 50 million); (4) the enhanced oil recovery tax credit ($0); and (5) marginal oil and gas production tax credits ($0).

Chapter 4 - The price of crude oil began to increase in the last quarter of 2003, and has led to the high prices observed from 2004 through 2007. The Iraq

War, unexpectedly high demand growth in China, India, and the United States, and Hurricanes Katrina and Rita, along with a number of other factors, all contributed to the rising price.

An important result of these largely unexpected events was that the oil industry, as represented by the five major integrated oil companies doing business in the United States, experienced rapidly expanding revenues and profits. Some observers characterized these profits as "windfall" gains, while others pointed to the increasing scarcity and rising costs observable in the oil industry. Some saw "price gouging" in high gasoline prices, while others saw the market working to avoid physical supply shortages.

The larger profits experienced by the oil industry and the five major integrated oil companies can be used in a variety of ways. Profits might be used to expand exploration and development of crude oil resources to expand the supply of oil. Refineries might be constructed, and technology improved at existing refineries, to expand the supply of petroleum products, most notably, gasoline. Profits might also be used to provide increased returns to the owners of the oil companies, the shareholders. This end might be accomplished through dividend payments and share repurchase plans. Profits might also be used to improve the balance sheets of the companies through debt reduction, potentially improving their financial health should they face a downturn in the market in the future. Until the effects of corporate plans that reflect a market characterized by higher oil prices can be observed, profits might tend to build up as cash reserves, as experienced by some of the five firms since 2004.

How the profits generated over the past four years are used will help to determine whether oil and petroleum product markets remain tight with high prices, or whether they loosen, develop extra capacity, and lead to moderating prices.

In: The Oil Industry and Windfall Profits
Editor: Maria B. Coswell
ISBN 978-1-60692-840-0

Chapter 1

THE CRUDE OIL WINDFALL PROFIT TAX OF THE 1980S: IMPLICATIONS FOR CURRENT ENERGY POLICY*

Salvatore Lazzari

SUMMARY

In April 1980, the federal government enacted the crude oil windfall profit tax on the U.S. oil industry. The main purpose of the tax was to recoup for the federal government much of the revenue that would have otherwise gone to the oil industry as a result of the decontrol of oil prices. Supporters of the tax viewed this revenue as an unearned and unanticipated windfall caused by high oil prices, which were determined by the OPEC (Organization of Petroleum Exporting Countries) cartel.

Despite its name, the windfall profit tax (WPT) was actually an excise tax, not a profits tax, imposed on the difference between the market price of oil and an adjusted base price. While most domestically produced oil was subject to the tax (about 2/3 in 1985), the remaining 1/3 that was tax-exempt was significant (1.3 billion barrels in 1985, or 360,000 barrels per day). The $80 billion in gross revenues generated by the WPT between 1980 and 1988 was significantly less

* This is an edited, excerpted and augmented edition of a CRS Report for Congress publication, Order Code RL33305, dated March 9, 2006.

than the $393 billion projected. Due to the deductibility of the WPT against the income tax, cumulative net WPT revenues were about $38 billion, significantly less than the $175 billion projected. This report presents estimates of the amount of foregone oil production from 1980-1986 due to the WPT under three alternative supply price responses, reflecting three different assumptions about the price elasticity of the domestic oil supply function, a critical factor (statistic) in estimating lost oil output and increased import dependence. From 1980 to 1988, the WPT may have reduced domestic oil production anywhere from 1.2% to 8.0% (320 to 1,269 million barrels). Dependence on imported oil grew from between 3% and 13%. The tax was repealed in 1988 because (1) it was an administrative burden to the Internal Revenue Service (IRS), (2) it was a compliance burden to the oil industry, (3) due to low oil prices, the tax was generating little or no revenues in 1987 and 1988, and (4) it made the United States more dependent on foreign oil. The depressed state of the U.S. oil industry after 1986 also contributed to the repeal decision.

Reinstating the windfall profit tax would reduce recent oil industry windfalls due to high crude and petroleum prices but could have several adverse economic effects. If imposed as an excise tax, the WPT would increase marginal production costs and be expected to reduce domestic oil production and increase the level of oil imports, which today is at nearly 60% of demand. Crude prices would not tend to increase. Some have proposed an excise tax on both domestically produced and imported oil as a way of mitigating the negative effects on petroleum import dependence. Such a broad-based WPT would tend to reduce import dependence, but it would lead to higher crude oil prices and likely to oil industry profits, potentially undermining its original goals. Because the pure corporate profits tax is relatively neutral in the short run — few, if any, price and output effects occur because marginal production costs are unchanged in the short run — a possible option would be a corporate income surtax on the upstream operations of crude oil producers. Such a tax that would recoup any recent windfalls with less adverse economic effects; imports would not increase because domestic production would remain unchanged. In the long run, such a tax is a tax on capital; it reduces the rate of return, thus reducing the supply of capital to the oil industry.

INTRODUCTION

Recent high crude oil and petroleum product prices, have sharply increased oil industry profits, which have prompted some policymakers to propose that either the windfall profit tax of the 1980s (WPT) be reinstated or that some new version of that tax be enacted. Currently, fourteen congressional bills propose some type of windfall tax on domestic oil companies. Three of these bills have been offered as amendments to S. 2020, the Senate's tax reconciliation bill. Such proposals are not uncommon when oil prices rise sharply to high levels. The doubling in oil prices from June to August of 1990, due to the crisis in the Middle East (Iraq invaded Kuwait on August 2, 1990), prompted similar proposals to reduce what many policymakers viewed as an unduly large increase in oil industry profits.

This report provides an overview and analysis of the crude oil windfall profit tax that was in existence from 1980 to 1988. The first section provides a brief history of events surrounding enactment of the tax. The second section describes the structure of the tax. The third section discusses the rationale for the tax. The fourth section examines the revenue effects and other economic effects, and shows how the windfall profit tax increased dependence on imported oil. The lessons that can be learned from the eight-year experience with the tax and the implications of reinstating the tax are discussed in the final section.

BACKGROUND

The Crude Oil Windfall Profit Tax Act (P.L. 96-223) was enacted in 1980 as part of a compromise between the Carter Administration and the Congress over the decontrol of crude oil prices. The structure of the WPT was based on the structure of the oil price control program.

The Oil Price Control Program

From August 1971 to January 1981 the price of domestic oil was controlled by the federal government — it was constrained from rising to market levels. Oil price controls were initiated under President Nixon's general wage-price freeze of August 15, 1971, which consisted of four stages or time periods called phases, each with various degrees of controls.[1] These controls were not focused on oil alone — they applied generally to most goods and services. The general wage-price freeze terminated in 1973, but under phase IV of the program — effective from August 12, 1973 to April 30, 1974 — price controls on petroleum were retained.

Oil price controls were extended through 1975 under the Emergency Petroleum Allocation Act of 1973 (P.L. 93-159) — enacted in November 1973, at the onset of the Arab oil embargo.[2] This law created the two-tiered price control program, making the structure more complex. Most domestically produced oil was categorized as either "lower tier (old) oil" or "upper tier (new) oil," each with its corresponding controlled base price. Lower tier oil was generally oil from properties that began production before 1973. Under regulations, the price ceiling on this oil was the highest posted price in effect on May 15, 1973 plus $0.35/barrel on all oil produced from wells that produced at less than their 1972 levels, which resulted in a posted price for old oil of about $4.25/barrel. Subsequent regulations increased the ceiling price in 1973 to $5.25/barrel. New oil (oil produced from wells that began production in1973), stripper oil (oil produced from wells that produce 10 bpd or less), and imported oil were not price controlled — it could be sold at the market clearing price, which was determined by the delivered prices of imported crude inclusive of the customs duties. Note that these tiers were artificial designations for the purposes of controlled pricing; the distinctions were made only for purposes of establishing different base prices; they had no other meaning.

Oil price controls were amended under the Energy Policy and Conservation Act of 1975 (EPCA75), and the Energy Conservation Act of 1976 (ECA). EPCA75 (P.L. 94-163) replaced the two-tiered price control system with a three-tier system. Lower tier oil was oil produced below the "base production control level," — the level below the 1975 average monthly production; upper tier oil was output in excess of this base level or output from new wells — those brought into production after 1975. Stripper oil was initially classified as (controlled) upper tier oil, but subsequent changes (see below) decontrolled it altogether. EPCA75 also established a national average price for all oil of $7.66/barrel, the weighted average price of lower tier oil (old oil), which had a ceiling of $5.25/barrel and upper tier oil (new oil and stripper oil), which, by this time also became controlled at $11.28/barrel. The average domestic oil price was permitted to increase at the rate of 10% annually. (In 1975, uncontrolled imported oil sold for about $13.25/barrel. At the onset of the oil price control program, domestic oil prices were just above $3 per barrel.) EPCA75 also gave the President the discretionary authority to end oil price controls anytime between May 31, 1979 and September 30, 1981 — on October 1,1981 all controls would expire. The ECA of 1976 made further (mostly technical) amendments to the price control system — e.g., exempting stripper oil from the price controls — but otherwise it left the basic structure of the oil price control program intact. Thus, in 1979, on the eve of the WPT, the oil price control program categorized all domestic oil production into these tiers, each with its own corresponding controlled or market price. Table 1 summarizes this structure, which was the foundation for the 1980 WPT.

EPCA75 also gave the President discretionary authority to gradually phase out oil price controls over the 28 month period between June 1, 1979, and September 30, 1981, after which prices would rise to world market levels. On April 5, 1979, the Carter Administration announced its intention to use that authority, and throughout its remaining term in office price controls on petroleum products and various types of crude were selectively decontrolled. The intent of the gradual decontrol was to promote energy conservation and to stimulate energy exploration and production without the dislocations that might result with sudden decontrol. Between April 1979 and January 1981, oil prices were partially and gradually decontrolled — newly discovered oil was completely decontrolled, but price controls on other types of oil continued. President Reagan repealed price controls on January 28, 1981, which was one of his first official decisions as President.[3]

Origins of the Crude Oil Windfall Profit Tax of 1980

Excess profits taxes are extraordinary measures as tax policy instruments, their use limited to wartime or other periods characterized by economic emergencies and instabilities such as inflations and hyper-inflations. Such was the case with the surtax on business profits imposed as a temporary measure to control large profits earned during World Wars I and II, and the Korean War.[4]

The origin of the 1980 WPT was, as with many other energy taxes and energy tax subsidies, the first oil crisis of the 1970s — the Arab Oil Embargo of 1973-74. Between 1973 and 1980, attempts to decontrol oil prices were accompanied by proposals to impose an excess profits tax. In 1974, the Ford Administration proposed an "emergency windfall profit tax" to recapture oil industry windfalls resulting from price decontrol. In August of 1975, the Senate Finance Committee approved a windfall profit tax conditioned on price decontrol. Congress, however, did not decontrol oil prices. In 1977, the Carter Administration proposed an oil tax similar to the 1980 windfall profit tax — the crude oil equalization tax — as part of a program to restructure the existing price controls.[5] These proposals were the precursors of the 1980 windfall profit tax.

The windfall profit tax that was ultimately enacted in 1980 originated with President Carter's April 5, 1979 decision to gradually phase out price controls between June 1, 1979, and September 30, 1981, as discussed above. At the same time, the Carter Administration announced its proposal to impose a WPT "to prevent unearned excessive profits" by the oil industry. In Congress, the Carter Administration's original windfall profit tax bill, H.R. 3919, was proposed by Representative Al Ullman, Chairman of the Ways and Means Committee. The House approved an amended version on June 28, 1979. The Senate approved its version at the end on 1979, and the House-Senate Conference deliberated for about three months. President Carter signed the bill into law on April 2, 1980.

Table 1. Summary of the Oil Price Control Program in 1979

Type of oil	Definition	Quantity of production (million barrels/day)	Price at the wellhead ($/barrel)
Lower Tier Oil (Old Oil)	Oil from properties that began producing before 1973	3.0	$5.86
Upper Tier (New Oil)	Oil from properties that began producing after 1973	3.0	$13.06
Uncontrolled Oil	Stripper oil, Alaskan North Slope oil, and oil from the Naval Petroleum Reserve	2.6	$18.50[a]

Source: U.S. Congressional Budget Office, *The Decontrol of Oil Prices: An Overview*, Background Paper, May 1979, pp. 5-7.

[a] Price for uncontrolled oil is at the refinery gate, which is basically the import price plus transportation costs.

Amendments to the Windfall Profit Tax

The WPT was amended many times after it was enacted in 1980. Virtually every tax law that was enacted between 1980 and 1988 made some type of amendment to it, usually minor and technical, but sometimes, significant. For example, the Technical Corrections Act of 1982 (P.L. 97-448) made several minor and technical amendments. Major changes were made, however, under the Economic Recovery Act of 1981 (P.L. 97-34), which reduced tax rates on newly discovered oil, made stripper oil (from small wells) tax exempt, and introduced a tax credit for royalty owners.

Soon after enactment of the WPT there were proposals to reduce tax rates, liberalize some of the provisions. Proposals to repeal the WPT altogether were also common throughout the eight-year life of the tax. Repeal was part of President Reagan's platform in the 1980 presidential campaign and repeal proposals were embodied in the Administration's FY1988 and FY1989 budgets.[6] WPT repeal was part of the Treasury Department's tax reform proposal of 1984 as part of a compromise that would have repealed the oil industry's two major tax incentives (subsidies): the percentage depletion allowance and expensing of intangible drilling costs. In the spring of 1986, there was speculation that a WPT repeal proposal would be part of the tax reform bills of 1986. The eventual law the Tax Reform Act of 1986 (P.L. 99-5 14) — did not, however, repeal the WPT.

Another major repeal attempt was made through an amendment to the 1986 debt limit bill (H.J.Res. 668) which would have increased the debt ceiling to over $2.3 trillion through FY1987. This amendment was approved by the House and Senate, but it was deleted in conference. Some Members of Congress also favored repeal; congressional support for repeal probably reached a peak in 1987 and 1988. In May 1987, for example, ten bills were pending in the Congress proposing to repeal the WPT.

Repeal of the Windfall Profit Tax

The actual repeal of the WPT in 1988 was made through an amendment to omnibus trade legislation (H.R. 3). After hearings in the summer of 1987, the Senate voted 58-40 in favor of a repeal amendment to the trade bill. The original House trade bill, however, did not contain the amendment. While House conferees were generally opposed to WPT repeal, conferees from the Senate Finance and House Ways and Means Committees did agree on a repeal amendment on March 31, 1988. The House approved H.R. 3 with the repeal amendment on April 21, by a vote of 3 12-107. The Senate approved the conference report on H.R. 3 on April 27. Enactment of H.R. 3 including the WPT repeal amendment was precarious because President Reagan — objecting to several of the provisions in the trade bill — had threatened to veto it. Eventually, however, these problems were resolved and the trade bill with the repeal of the WPT was signed in August of 1988.[7]

Structure of the Tax

Despite its name, the crude oil windfall profit tax was not a tax on profits. It was an excise tax — or more accurately, a system of excise taxes — on domestically produced oil effective March 1, 1980. The tax was imposed on the difference between the market price of oil, which was technically referred to as the removal price, and a statutory 1979 base price that was adjusted quarterly for inflation and state severance taxes.[8]

All domestically produced oil that was not specifically tax-exempt was classified into one of three categories or "tiers" based upon the age of the well, the type of oil, and the amount of daily production. These categories were a carryover from the oil price regulations which also categorized oil into various tiers. For example, tier I oil was oil classified as lower and upper tier oil under price controls; tier II and tier III oil was oil exempt from price controls. Domestic crude oil is oil produced from an oil well located in the United States, or in any of its possessions, and it includes the Continental shelf area of the United States.

For each oil category there was a corresponding tax rate (or rates, as explained below) and a corresponding adjusted base price. The tax rates and adjusted base prices differed not only according to the type of oil but also according to whether an oil producer was an integrated producer (called a major) or an independent oil producer. The tax rates applicable to oil sold by an independent oil producer were lower than the tax rates applicable to oil sold by a major integrated producer.[9] The windfall profit tax liability on any barrel of oil was limited to 90% of the net income (profit) from the sale of that oil, which was basically the taxable income (gross income less costs) corresponding to that same oil.

Five categories of oil were originally exempt from the windfall profit tax: (1) oil owned by a state or local government or any political subdivision thereof; (2) oil owned by a qualified educational institution or a charitable medical institution; (3) oil owned by Indian tribes or individual Indians on January 21, 1980, over which the U.S. exercises trust responsibilities; (4) new oil produced from much of Alaska; and (5) "front-end tertiary oil."[10] Stripper oil, and a limited amount of oil produced by royalty owners, became tax exempt under an amendment in the Economic Recovery Tax Act of 1981. Table 2 shows the structure of the tax just prior to its repeal. Note that newly discovered oil is oil produced from a well which was not producing in 1978. For perspective, the average world market price was $32.37 in 1980, and $13.25 in 1988.

The WPT was imposed on domestic oil producers at the point of the first sale of taxable oil, which was generally to a refiner. The refiner — known also as the first purchaser — withheld the tax from the amounts otherwise payable to a producer and deposited the funds semi-monthly into an account. In other words, the tax amount per barrel was subtracted from the oil's purchase price. The first purchaser was required to file tax returns on a quarterly basis. In cases where withholding was not required, such as when the producer and first purchaser were one and the same, the tax was paid directly to the Treasury.

The WPT was a deductible expense in determining an oil producer's income tax liability because, as with all other excise taxes, it was considered a cost of doing business. As explained in a later section, this meant that, as a result of paying the WPT, a producer's income tax liability was lower than it would have been without the tax..

The WPT was a temporary tax. The statute provided that the tax would *begin* to phase out sometime during the three-year period between January 1988 and January 1991. The precise starting point for the phase out depended on cumulative net revenues. If the Secretary of the Treasury reported that, on a given month beginning January 1988, cumulative net revenues would exceed the pre-established target of $227.3 billion, then the phaseout of the tax would begin on the month following the attainment of the target. If estimated cumulative net revenues would not exceed $227.3 billion between January 1988 and January 1991, then the phase out of the tax would begin January 1991. Irrespective of the onset of the phaseout, once the phaseout began, the tax was to have been phased out over a 33-month period by reducing each producer's tax by 3% each month (with 4% on the last month).

Cumulative net revenues, which totaled about $43 billion between 1980 and 1987, fell far short of the $227.3 billion target. Hence, had it not been repealed, the WPT phase out would have begun January 1991 and the tax would have terminated on October 1, 1993. As already discussed, the WPT was repealed in August 1988, before the onset of the phase-out rules.

Finally, while the tax was called a "profit" tax, it was not really a profit tax but rather a special type of excise tax — a selective excise tax on oil producers. The tax was paid first, before profits from the sale of the oil were determined. And except for the net income limitation, profits had no bearing on how much WPT was paid. The base prices had no precise or even approximate relationship to the costs of oil production. This difference between an excise tax and a true excess profit tax is crucial because, as will be demonstrated, the two taxes have very different economic effects, particularly on energy prices and oil imports.

Table 2. Structure of the Crude Oil Windfall Profit Tax

Oil Type (1)	**Tax Rate (2)**	**Average Base Price in 1980 ($/barrel) (3)**	**Average Base Price in 2nd Quarter of 1988 ($/barrel) (4)**
Tier I (most domestic oil in reservoirs productive before1979)	70% for Majors 50% for Independents	$12.81 $12.81	$19.54 $19.54
Tier II (oil from stripper wells and the Naval Petroleum Reserve)	60% for Majors 30% for Independents	$15.20 $15.20	$23.19 $23.19
Tier III (heavy oil, incremental tertiary oil, and newly discovered oil)	30% for heavy oil and incremental tertiary oil 22.5% for newly discovered oil	$16.55 $16.55	$22.92 $22.92

Source: Sections 4986-4998 of 1986 Internal Revenue Code; Commerce Clearinghouse, 1987; and Research Division of the Internal Revenue Service.

Notes: Tier I oil is oil that was lower and upper tier oil under EPCA75, and Sadlerochit oil from Alaska's North Slope; Tier II and Tier III oil was uncontrolled oil under EPCA75; The 1981 ERTA legislation changed the rate structure as follows: (1) The tax rate on new oil was reduced gradually, but further reductions were frozen in 1984 so that the rate remained fixed at 22.5% until 1988; (2) Stripper oil produced by independents was made tax-exempt; (2) Royalty owners received a tax credit followed by an exemption for limited amounts of oil production.

RATIONALE FOR THE WINDFALL PROFIT TAX

The WPT resulted from a compromise between the Carter Administration, which wanted to decontrol oil prices, and the Congress, which generally did not; without the tax, many doubted that the Congress would have supported oil price decontrol.

The reasons for the tax are manifold and complex; they transcend economics and they concern the image and perception of the oil industry. The record does show, however, that the Congress was concerned that the industry would reap enormous revenues and profits as a result of decontrol to world oil price levels.

The Congress believed that the projected huge redistribution of income from energy consumers to energy producers would not be fair. The Congress was concerned that the oil industry was not paying its fair share of federal taxes. And finally, the Congress was looking for additional sources of revenue.

Oil Price Decontrol and Windfall Profits

Price decontrol implied that domestic crude oil prices would rise from an average of about $14 per barrel (1979) to world market levels, which at that time were averaging about $24 per barrel and projected by some to rise to $50-$60 per barrel or more by 1985.[11]

In 1979, many analysts were predicting sharp increases in domestic oil prices and, with them, significant increases in oil industry revenues and profits. The Joint Committee on Taxation had estimated that decontrol would increase oil industry revenues by about $1 trillion from 1980-1990 and profits by over $400 billion. Federal policymakers believed that these added profits were in the nature of a "windfall" — an unearned, unanticipated gain in income through no additional effort or expense. This windfall was thought to provide no additional incentive to produce more oil, especially "old oil," which was already being produced under the preexisting controlled price regime. Rather, existing oil reserves would simply be worth more — command a higher price by virtue of price decontrol. Moreover, all oil would command a higher price, including oil that was discovered at historically low costs, and produced at the controlled price. Old oil, which was primarily owned and produced by the major oil companies, was selling for about $6 per barrel prior to decontrol; after decontrol it would have increased to a market price of about $24 per barrel. The Congress was concerned that no additional effort, investment, or cost would be incurred by oil producers in generating the added profits. The Congress also believed that a higher price was not needed for all oil in order to stimulate its production — but that a higher price might be needed for new oil. The decision to produce much of the oil had been made with the expectation of a return based on the controlled price.

The following quote illustrates the concerns of the Congress:

> For most types of oil, after a certain point, these higher prices will only lead to very limited increases in production. The revenues resulting from these higher prices, however, would provide income to oil producers far in excess of what most of them originally anticipated when they drilled their wells and in excess of

> what they might now be expected to invest in energy production. Indeed, some producers are now using their excess revenues to acquire unrelated businesses.
>
> Thus, the committee believes that the additional revenues received by oil producers and royalty owners, both as a result of decontrol of oil prices and as a result of increases in world oil prices substantially above those prevailing in 1978, are an appropriate object of taxation. The windfall profit tax in this bill will tax away a fair portion of these additional revenues while allowing producers to receive very high prices for those types of oil whose production can be expected to increase in response to that incentive.[12]

Other motivations and factors underlying imposition of the windfall profits tax were that

- Domestic crude prices would rise to market levels that did not reflect competitive market forces but the market power of OPEC; and, further, OPEC's prices were projected to increase in real terms at very high rates, usually assumed to be 3% per year.
- The market price of oil was believed to be in a sense "unanticipated," unearned, and unneeded for the profitability of the oil industry.
- Society should share in the economic return to natural resource production.
- Oil is a natural resource whose long-run supply is fixed; it is not like other factors of production such as labor and capital. The stock of natural resources is fixed in the long run whereas the stock (or supply) of the other factors is variable. Since the stock of oil is fixed, some argued that high levels of industry income were not necessary to ensure adequate supplies. If low levels of income would ensure adequate oil supplies, then any industry income above that income earned from alternative use of industry resources could be deemed excessive (economic rents) and should be taxed away.
- Some believed oil industry income was excessive to start with due to the concentrated structure of the domestic oil industry and due to the fact that domestic price of oil was not a competitively determined price.

Additionally, it should be remembered that the WPT was enacted in the wake of two oil shocks in the 1970s: (1) the 1973-1974 oil embargo, which raised oil prices fourfold and (2) the 1978-1979 Iranian revolution, which doubled oil prices, and created gasoline shortages (and long lines of motorists at the gasoline pumps). Also there was a certain amount of public suspicion of the oil industry;

suspicion that the energy crisis was not real but a contrivance of the industry in concert with OPEC for the purpose of profiteering.

Distributional Equity ("Fairness")

Another rationale for the windfall profit tax was equity or "fairness." It was estimated that oil price decontrol would cause a large redistribution of income from energy consumers to energy producers. Policymakers concluded that it was unfair for the oil industry and landowners to experience such sharp increases in income when so many consumers — particularly low-income consumers — would see a sharp increase in their energy bills. They believed that society at large, through the federal government's policies, should also share in some of the income gains.

The fairness rationale was strongly influenced by the impact of higher energy prices on poorer consumers. Although all energy consumers would experience a higher absolute burden due to higher oil prices, including higher electricity prices, natural gas prices, and coal prices, poorer people would experience a higher relative burden. That is, in relation to their income, poorer persons spend more money on energy and other necessities than higher income persons. Therefore, energy costs represented a higher proportion of low income persons' budgets than high-income persons' budgets — the burden from decontrol would be greater for low-income persons than from high-income persons.

The windfall profit tax was intended to be the instrument for achieving a more equitable redistribution of the income which would result from oil price decontrol. Underlying this instrument was the belief that the oil companies were entitled to a fair and reasonable return but not an "excessive" return, which was in any event determined by OPEC-set prices rather than competitive prices.

The Industry's Historically Low Effective Tax Rates

Another powerful argument for enacting the WPT was that the tax helped to offset the oil industry's low effective income tax rates due to the availability of two oil industry tax subsidies (incentives): the percentage depletion allowance and the provision which permits companies to expense (deduct fully in the initial year) the intangible costs of drilling.

The percentage depletion allowance permits oil producers to deduct an amount for the exhaustion of an oil reserve equal to a percentage of revenues. In theory, the deduction should be based on the actual oil output and the actual investment costs of the deposit — it should be cost depletion. The percentage depletion allowance was introduced in 1926. In 1975 the allowance was eliminated except for a limited amount of oil produced by independents. The deduction for intangible drilling costs permits oil producers to expense — deduct contemporaneously — costs that, according to economic theory and standard financial accounting practices, should be capitalized over the income-producing life of the deposit. This subsidy or incentive was introduced in a 1918 administrative ruling by the Treasury Department. The cumulative value of these tax and other nontax subsidies from 1964-1977 has been estimated at over $100 billion.[13]

In recent years, the value of these oil and gas subsidies has declined,[14] but the addition of other tax subsidies, such as the enhanced oil recovery tax credit, and, more recently, the oil and gas tax breaks in H.R. 6 (P.L. 109-58) have, at a time of very high oil prices, created more support for a windfall profit tax.[15]

The combined effect of the two major oil tax provisions was to lower effective income tax rates for oil extraction below the comparable effective tax rates in other industries and below the top marginal statutory income tax rate of 34% for corporations in 1980. This is supported by early as well as more recent empirical research studies on effective tax rates.[16]

In the early studies, Harberger (1955) and Steiner (1959) demonstrated that oil and gas, as well as other minerals, received approximately twice the amount of tax incentives as other industries. In the category of effective tax rate studies, a 1971 report by U.S. Oil Week showed that major oil companies had an effective tax rate of 8.7% in 1970. Cox and Wright (1973) calculated rates ranging from 8.3% to 14.7%, depending upon accounting methods and income measures used.[17]

Studies on effective tax rates published between 1973 and 1980 attempted to include the cutback in subsidies and the windfall profits tax and gave mixed results. Some studies, for example, showed that oil and gas extraction was subject to very low effective tax rates. Several studies by the Congressional Research Service published between 1977 and 1983 (when the corporate tax rate was 46%) show very low and, under certain circumstances, even negative marginal effective tax rates. For example, expensing of intangible drilling costs and dry hole costs and a 22% depletion rate resulted in an effective tax rate of -3.0% without the minimum tax and 12.0% with the minimum tax.[18] One CRS report, which included the effects of the crude oil windfall profits tax, again showed generally low effective tax rates for oil and gas extraction. In cases where the effective tax rates were low, however, the crude oil windfall profits tax constituted a significant part of the total effective tax burden.[19] In an inter-industry comparison, oil extraction and production had the lowest effective tax rates of eleven major industries — 14% compared to 17% for construction (the next lowest) and 30% for the trade industry (the highest).[20]

Marginal effective tax rates in the oil and gas industry after 1986 increased due to the repeal of the 10% investment tax credit, the lengthening of the recovery period for depreciation, and the change in the depreciation methods. Studies continue to show, however, marginal effective tax rates below the statutory top marginal tax rate, and below the comparable rate for most other industries.[21]

A 2000 Institute on Taxation and Economic Policy (ITEP) study found that, for the three-year period 1996-98, petroleum and pipeline companies had the lowest effective tax rate (12.3%) of 20 industries (the average effective tax rate for all the industries was 21.7%).[22] The study sampled 250 of the nation's largest corporations. In a 2004 update of the 2000 study, ITEP found petroleum and pipeline companies had effective federal tax rates of 13.3% for the three-year period 2001-2003, but were no longer ranked number one of the 20 industry classifications — they were ranked 6th.[23] Gruber (2005) found that the mining and extraction industries combined had an average marginal effective tax rate of 16.8%. While this was significantly below the 35% marginal statutory tax rate for the period, it was only slightly below the industry average of 17.4%.[24]

Budget Deficits and the Need for Additional Revenue

There were also important fiscal reasons for enacting the WPT — the federal government needed money. Between 1961 and 1979 the federal budget was in deficit in every year but one (there was a small surplus in FY1969). In FY1976 the deficit reached $71 billion, which at that time was the highest level in U.S. history. This deficit was 4.2% of Gross Domestic Product (GDP) the highest since 1946. In fiscal years 1977, 1978, and 1979, the deficits were lower but still sizeable — $50 billion, $55 billion, and $38 billion, respectively. Certainly they pale in comparison to the deficits of the eighties and early 1990s but according to the standards of that time they were still large. Recent deficits have been larger in absolute terms but somewhat smaller relative to GDP. For example, in FY2004, total federal budget deficits was $413 billion, which was 3.6% of GDP.[25] In any event, revenues and deficit reduction have not been the driving force behind the recent WPT proposals in the Congress, although with large federal spending on the Iraq war, and hurricane relief, any additional revenues would, if not spent, reduce the deficit.

Economic Effects

The major economic issues concerning the WPT and its effects were: revenues, increased dependence upon foreign oil, economic efficiency, and the tax's administrative and compliance burden. As discussed above, the need for revenue was one of the reasons for enacting the windfall profit tax and was a principal issue in the debate over its repeal. However, the tax's role in increasing dependence on imported oil, distorting resource use in the energy markets and the economy, as well as the administrative and compliance burden of the tax, all played a role in its repeal.

Revenue Effects: Projections vs. Realizations

Table 3 compares the original projections of gross windfall profit tax revenues with actual revenues for fiscal years 1980-1990. Gross revenues are the actual tax monies collected by the Internal Revenue Service (IRS) as a result of applying the WPT rates to taxable crude oil production — they are revenues before any deductions or allowances, and offsets. Note that these original 1980 projections are not adjusted downward for changes in the tax laws enacted in 1981, 1982, and 1984, which tended to reduce windfall profit tax revenues.[26]

As these data show, estimates or *projections* of the additional tax revenues from decontrol with the WPT indicated that the federal government would generate, over the 11-year period between fiscal years 1980-1990, an additional $393 billion in gross revenues. Net revenues for this period were projected at $223 billion, reduced by the loss of $170 billion in business income tax revenues due to the deductibility of excise taxes as a cost of doing business. Including state and local severance taxes and income taxes, and taxes on royalty income, all levels of government were projected to receive about 50% of the additional revenue from oil price decontrol. The oil industry was projected to receive the remaining 50%.[27]

Large overestimates of projected revenues occurred in the original forecast of revenues made in 1979 and 1980, reflecting overestimates of crude oil prices.[28] The decline in oil prices in the mid-1980s was not anticipated even in 1981. In fact, it was after 1983 that analysts began to adjust their oil price forecasts downward in consideration of new sources of oil supply, increased conservation of oil, and the development of alternative energy resources.

Table 3 also shows that, for the same period, actual gross revenues were about $80 billion, significantly short of projections — 80% less than the projected amount of $393 billion — but still a sizeable sum.[29] Most of this gain was accumulated over the years 1981-1983, when gross revenues totaled nearly $55 billion. These large initial revenues from the windfall profit tax were also an important reason in the early part of the 1980s for not repealing the tax, despite President Reagan's campaign promise and numerous congressional attempts to repeal it.[30]

Table 3. Gross Windfall Profits Tax Revenues: Projected vs. Actual, Fiscal Years 1980-1990 ($ millions)

Year	Projected tax revenues (1)	Actual tax revenues (2)	Projected less actual (1) - (2) – (3)	Actual as % of projected (4) = (2)/(1)
1980	5,159	3,052	2,107	59%
1981	20,955	16,931	4,024	81%
1982	32,293	22,036	10,257	68%
1983	35,124	15,660	19,464	45%
1984	37,129	8,120	29,009	22%
1985	39,535	5,073	34,462	13%
1986	40,923	8,866	32,057	22%
1987	52,524	15	52,509	0%
1988	44,181	373	43,808	1%
1989	46,270	30	46,240	0%
1990	48,538	(86)	48,624	N.M.
Total: 1980-1990	**392,931**	**80,070**	**312,861**	**20%**

Sources: (1) Projected figures are from U.S. Congress. Joint Committee on Taxation. *General Explanation of the Crude Oil Windfall Profits Tax Act of 1980* (H.R. 3919, 96th Congress; P.L. 96- 223). Joint Committee Print. Washington, 1981. p. 15; (2) Actual tax revenues for FY1980-FY1986 are from quarterly excise tax reports published by the Internal Revenue Service. Data for 1986-1991 were obtained from the Congressional Budget Office.

N.M. = not meaningful.

Table 4. Estimated Revenue Effects of the WPT, Fiscal Years 1980-1 990 ($ millions)

Year	Gross Revenues (1)	Less Receipts from Federal Interests (2)	Less Estimated Foregone Income Tax Payments (3)	Equals Estimated Net Revenue Effect (4)
1980	3,052	492	1,404	1,156
1981	16,931	1,105	7,788	8,038
1982	22,036	1,092	10,137	10,807
1983	15,660	902	7,203	7,555
1984	8,120	757	3,735	3,628
1985	5,073	601	2,334	2,138
1986	8,866	567	4,078	4,221
1987	15	1	6	8
1988	373	N.A.	127	246
1989	30	N.A.	10	20
1990	(86)	N.A.	N.A.	N.A.
Total: 1980-1990	**80,069**	**5,517**	**36,685**	**37,817**

Sources: Column (1)data are from Francis, Brian. *Federal Excise Taxes, Including the Slow Death of Expired Taxes.* Internal Revenue Service. Statistics of Income Bulletin, summer 1999. pp. 185- 189. Note that 'Gross Revenues' includes adjustments made as a result of errors, IRS and court challenges, and other factors; column (2) data are from Hakken, John. *Windfall Profit Tax Liability and Receipt Estimate.* U.S. Treasury. Office of Tax Analysis; column (3) is calculated by the author based on the marginal statutory corporate tax rate at the time (46% except for 40% in 1987 and 34% in 1988.

N.A. = not available.

Finally, note also that after FY1982, gross revenues (column [2]) began to decline, sharply beginning in 1984, and down to nearly zero in 1987. There are three reasons for this. First, market crude oil prices declined markedly from 1982 to 1986. Second, since 1980 base prices had been gradually adjusted upward due primarily to inflation, as specified by law. The result was two forces acting to reduce the tax base — the so-called "windfall profit." The third reason was the decline in domestic oil production. As will be discussed in a forthcoming section, the small amount of revenue collected from the WPT in 1987 and 1988 was a principal reason for the repeal of the tax.

Revenue Effects: Gross vs. Net Actual Revenue

Table 4 shows the estimated net WPT revenues, after adjusting for income tax offsets, and receipts from federal oil interests. Column (1) is the same as column (2) in Table 3. Column (2) shows WPT payments on federally owned oil attributable to the economic interests of the U.S. Government — WPT assessed on oil produced from federal lands. These figures are included in the gross WPT liabilities reported in column (1) because, under the law, oil produced from federal properties was not tax-exempt. In effect, then while the federal government was collecting WPT revenues on its properties, it was also receiving equally less in price because first purchasers subtracted the tax from the purchase oil price. Column (3) shows estimated foregone payments of income taxes — reduced income tax collections — due to the deductibility of the WPT against the income tax liability as a cost of production. Net revenues are shown in column (4).

The estimated net revenue gains from the WPT — the amount which actually went into the Treasury's general fund — were nearly over $38 billion, less than of gross revenues, and 17% of the net revenues predicted in 1980 over the 1979-1990 period.[31] Most of the difference between gross and net revenues was attributable to losses in business income taxes (both individual and corporate) due to the deductibility of the gross WPT payments. Between 1980 and 1989, as shown in column (4), income tax revenues were estimated to be about $37 billion lower as a result of the deductibility of the WPT. The remaining revenue losses were due to receipts from federal interests (column [2]).

Effects on Domestic Oil Production and Oil Imports

The WPT had the effect of reducing the domestic supply of crude oil below what the supply would have been without the tax. This increased the demand for imported oil and made the United States more dependent upon foreign oil as compared with dependence without a WPT. Nevertheless, oil price decontrol, by increasing prices should have increased domestic production and made the U.S. less import dependent. And further, while a WPT made the U.S. more dependent on imported oil, decontrol and a WPT made the U.S. less dependent than controls without WPT.

The WPT was a excise tax on oil produced domestically in the United States; it was not imposed on imported oil. In economic terms, such taxes increase marginal production costs, and profit maximizing firms respond to the tax by reducing output and raising prices. The WPT increased the marginal or incremental cost of domestic oil production subject to the tax — every barrel of oil produced cost more to produce by the amount of the tax. However, in the case of domestic crude oil, the higher marginal costs are not to be shifted as higher oil prices, because, oil being priced in the international (world) oil market — oil prices are exogenous to the U.S. (the U.S. is a price taker, rather than a price setter).[32] Oil producers could not shift the tax forward as a higher oil selling price because the purchaser would merely substitute imported or tax-exempt crude. Instead, the WPT reduces the net selling price paid to producers. As noted earlier, the first purchaser (generally the refiner) subtracted the tax from the price paid to the producer (supplier) — the producer's net selling price of each barrel of oil was less by the amount of the WPT.

This inability to shift the tax forward implies that the entire effect of the tax is to reduce domestic production and supply. In other words, U.S. domestic oil production was, to some degree, lower as a direct result of the WPT. But, as oil imports to the United States are a residual, the difference between aggregate demand for oil and aggregate domestic oil supply, the effect of this is an increase in the demand for oil imports.[33] Another way of stating this point is that imported oil is the marginal source of oil — whenever an extra barrel of oil is needed to meet an increase in demand, it is imported. Any condition or factor which either reduces domestic supply (such as higher industry taxes) or which increases the aggregate demand for oil (such as higher national income) will increase oil imports. and therefore also reduced the supply of the taxed product.

Estimates

The magnitude by which the WPT reduced domestic oil supplies and increased imports depends on two variables: (1) the magnitude of the decline in the supply price of domestic crude oil (the amount of the WPT), which determines the after-tax price received by oil producers; and (2) the price elasticity of the supply curve, which determines the reduction in oil production in response to the lower price (net of the WPT) received by oil producers. Since oil prices are determined in a world market, it is assumed that the WPT had no effect on pre-tax oil prices in the United States. This means that the after-tax price received by domestic oil producers is lower by the full amount of the WPT per barrel. (That is, as noted above, the WPT cannot be shifted forward in higher prices; producers absorb the entire tax in terms of lower profits.)

The second variable that determines the output effects is the price elasticity of the oil supply curve, which measures the responsiveness of oil production to changes in oil price.[34] An elasticity of + 1.0 means that a 10% reduction in the net price of oil to the producer translates into a 10% reduction in the quantity of oil supplied; an elasticity of +0.5 means that a 10% reduction in price would reduce output by half that or 5%. The price elasticity of oil supplies is determined by the technology underlying domestic oil production.

There is little doubt that crude oil production is relatively inelastic in the short run — even large price increases are unlikely to elicit substantial increases in production. Over the long-run however, producers can increase their investment and capital to increase production (development wells) from existing fields and increase exploration of new fields. The percentage reduction in oil production in response to the WPT would be the product of the percentage reduction in the after-tax price of oil times the price elasticity of supply. Oil supply price elasticities are difficult to estimate; few studies generate reliable estimates and in fact some studies estimate negative supply elasticities, which are not plausible. In developing our assumptions about the price elasticity of the crude oil supply curve we surveyed 20 studies.[35] In general, studies from the 1970s and early 1980s report larger supply price elasticities ranging from .1 to .8, with a mean of about .5.[36] Studies from the mid-1980s and more recent studies tend to show smaller supply elasticities. For example, a 1985 study for the Department of Energy found a convergence around a number ranging from .2 to .4; it used a value of 0.31 in its calculations.[37] Reflecting this uncertainty production losses are estimated under three alternative elasticity scenarios: a price elasticity of oil supply of +0.2 (short run), +.5, and +.8.

Table 5 presents estimates of annual domestic oil production that was lost in response to the WPT based on conventional assumptions. Note that unlike Tables 3 and 4, data are presented in calendar years to conform with the reporting of production and tax data. Estimates were prepared for the period 1980-1986. From 1986-1988 there are no output effects because the WPT liability was zero in those years. (The WPT was zero because market crude oil prices were below inflation adjusted base prices.) For perspective, the annual estimates of production losses are compared to the actual levels of domestic oil production and imported oil.

Table 5. Estimated Reduction in Domestic Oil Production in Response to the Windfall Profit Tax

Year	$\epsilon_p^s = +.2$			$\epsilon_p^s = +.5$			$\epsilon_p^s = +.8$		
	Million barrels (1)	**% of Total domestic output (2)**	**% of Imports (3)**	**Million barrels (4)**	**% of Total domestic output (5)**	**% of Imports (6)**	**Million barrels (7)**	**% of Total domestic output (8)**	**% of Imports (9)**
1980	74.0	2.0	3.9	184.0	5.0	9.6	294.0	7.9	15.3
1981	77.0	2.1	4.8	194.0	5.2	12.1	310.0	8.3	19.3
1982	58.0	1.6	4.6	145.0	3.9	11.4	232.0	6.2	18.2
1983	41.0	1.1	3.4	103.0	2.7	8.5	164.0	4.4	13.5
1984	36.0	1.0	2.9	101.0	2.6	8.1	161.0	4.2	12.9
1985	35.0	0.9	3.0	71.0	1.8	6.1	114.0	3.0	9.8
1986	0.0	0.0	0.0	0.0	0.0	0.0	0.0	0.0	0.0
1980-1986	320.2	1.2	3.2	794.5	3.0	8.0	1,268.8	4.8	12.7

Source: Author's estimates based on data published by the Department of Energy and the Internal Revenue Service.

Note: E^Ps denotes price elasticity of domestic oil supplies. This measures the responsiveness of domestic oil supplies to changes to the domestic price of oil. For example, an elasticity of +0.5 means that as the price of oil increases by 10%, the quantity of oil supplied increases by half that or 5%.

These estimates indicate that the windfall profit tax caused domestic oil production losses in every year but 1986, when crude prices declined below adjusted base prices resulting in zero WPT. Over the entire 1980-1986 period, it is estimated that, depending on the assumed supply curve price elasticity, the WPT reduced domestic oil production from between 320 million barrels (1.2% of domestic production) and 1,268 million barrels (4.8% of domestic production). The effect of reducing domestic oil production was to increase the level of imported oil. Columns (3), (6) and (9) show the estimated production losses caused by the WPT, as a % of the actual level of imported oil, under the assumed three supply curve elasticities range from 3.2% of total imports to 12.7% of imports for this period, depending on price elasticity.

If lag effects are discounted, the largest output effects were in 1981. That year the estimated loss in domestic oil production in response to the WPT ranged from 77 million barrels (2.1% of total domestic output) to 310 million barrels (8.3% of output), depending upon the assumed price elasticity. As a fraction of total imports, these constituted 4.8% and 12.1%, respectively.

Estimated annual production losses declined steadily between 1981 and 1986. This was due to the combined effect of declining market prices and increasing base prices over this period. In 1986 production losses were estimated to be zero because average market oil prices were below average base prices (the average windfall profit was negative in each of these years). It is important to note that the estimates in Table 5 assume that the production losses occur in the same year as the tax increase. In reality there may be lags in the effect of the WPT on domestic oil production. To this extent, the aggregate production losses estimated over the 1980-1986 period are probably more meaningful than the losses estimated for any one year.

On this basis the WPT increased oil imports and made the United States somewhat more vulnerable to sharp oil price increases or complete oil supply embargoes from foreign oil producers. If accurate, these estimates suggest that oil production losses in response to the WPT may have been from 3% to 13% of total oil imports.

Economic Efficiency and Resource Allocation

The efficiency effects of the WPT on the allocation of resources are less clear than some of the other economic effects. From an economic perspective, excise taxes distort the price system's ability to efficiently allocate resources among competing economic sectors. But the windfall profit tax had little if any effect on oil prices simply because such a tax cannot be forward shifted i.e., producers are not able to pass the tax forward by increasing prices to refiners because refiners would merely substitute imported oil. The reason for this is that oil prices in the United States are a given — they are determined or established in the world oil market in which the United States is only one of many producers.

In the long run, a permanent excise tax reduces the rate of return in the taxed sector and resources are allocated toward the non-taxed sectors. But the WPT was a temporary tax when it was enacted, and it was repealed two and one-half years before expiration. It is difficult to say whether this is a long enough period of time to cause resources to be reallocated in any significant way.

The other complicating factor was the decontrol of oil prices. It is a fundamental economic law that, generally, price controls cause serious distortions and create allocational inefficiencies. Oil price decontrol on January 28, 1981, when the WPT was about 10 months old, completely removed these distortions and inefficiencies. In all probability, decontrol with a WPT was less distorting than controls without a WPT.

The efficiency effects of the WPT also hinge on the question of whether oil production creates costs — these are called external costs — that society incurs but that producers do not account for. If the business of producing oil domestically creates these external costs, then an economically appropriate policy would have imposed some type of excise tax on oil production. The WPT that was in effect was not inconsistent with this policy and, to this extent, may have contributed to economic efficiency.

The WPT, however, may have distorted the way resources were allocated within the oil industry. Since the tax was imposed on oil production — i.e., upon its removal and sale — extraction (and other upstream operations) was penalized and other aspects of the business (refining and marketing, the downstream operations become relatively favored. Thus it created financial incentives to shift resources from exploration and drilling to refining and marketing.[38]

There may have been additional distortions within the oil-producing sector as a result of the structure of the tax. Under WPT, different tax rates and base prices applied to taxable oil, depending upon its classification in one of three tiers as described in Table 1. These differences seemed to favor oil from newer wells as opposed to oil from older wells, and oil produced from small wells and by independents, as opposed to oil produced from larger wells and by integrated producers. While this was probably done to minimize the adverse effects on supply, the structure of the WPT created artificial tax incentives based on the age and infrastructure of production and who owned the oil.

In addition to the above distinctions, the following categories of oil were tax exempt: (1) oil produced from a property owned by a state or local government or any political subdivision of a state government; (2) oil produced by educational institutions or charitable medical institutions; (3) oil produced from wells in certain regions of Alaska; (4) oil owned and produced by certain American Indian tribes; and (5) front-end tertiary oil and royalty oil.

Even so, it must be underscored that the distortions under the windfall profit tax with decontrol were probably less than the distortion under full price controls without a WPT.

The Burden of Tax Compliance and Administration

After 1986, the WPT imposed little or no tax liability on oil producers because oil prices were below the threshold base prices that triggered it. Oil producers were obliged to comply with the paperwork requirements of the law, however, and the Internal Revenue Service (IRS) was compelled to administer the system despite the fact that the tax generated no revenue.

The oil industry maintained all along that the WPT was an extremely complicated tax to comply with and to administer. The IRS and the General Accounting Office (GAO, now the Government Accountability Office) both agreed with the industry's claim, and the eight-year experience with the tax also tended to support this.[39]

The process of complying with the WPT involved a complicated system of interactions between a variety of oil industry entities and a variety of separate tax laws and energy regulations. The windfall profit tax was imposed on oil producers when taxable crude oil was removed from the oil-producing property. Any individual or business with an economic interest in an oil-producing property was considered as a producer and subject to the tax. There were four kinds of producers — independent producers, integrated oil companies, royalty owners (landowners), and tax-exempt parties. According to a 1984 GAO report, there were about 1 million oil producers (persons, institutions, and businesses) in the United States in 1984.[40]

Operators were the approximately 18,000 persons in the business of managing oil properties. The property operator supplied the relevant information to the agent who withheld the tax. The operator had to determine the proper tier, how much oil was sold, and who had the economic interest. Sometimes there were hundreds of people having a fractional economic interest in a single oil-producing property. Even determining the proper tier was no easy task. According to a 1982

GAO report, considerable uncertainty surrounded the concept of oil property, thus making it difficult to classify oil into tiers.[41]

The withholding agent had to compute and withhold the windfall profit tax based on the information supplied by the operator. The withholding agent, also called the first purchaser, was usually an integrated oil company, but it could also have been an independent producer or refiner.

To compute the windfall profit tax amount, the agent subtracted from the removal price the base price and the corresponding state severance tax (if any). This computation required the following steps: (1) knowing the category of oil; (2) determining the removal (selling) price; (3) adjusting the corresponding base price; (4) subtracting the state severance tax; and (5) testing for the 90% net income limitation.[42] Even some of the basic steps in this computation could be complex. For example, in 1983 there was some controversy over how to determine the "removal price" in the case of certain Sadlerochit oil in an Alaskan North Slope reservoir. Three different methods were used by the oil companies. The IRS had to issue several rulings before the matter was settled.

Having computed the tax liability, the first purchaser deducted this from the purchase price to be paid to the operator, and deposited the money in a Federal Reserve Bank. Integrated producers were required to deposit twice per month; independent producers were required to deposit every 45 days. The tax payment process did not, however, end there. In the event of overpayment or underpayment, due primarily to the net income limitation and underwithholding respectively, this required either refunds or additional payments.

Throughout this compliance process many tax return forms and information forms were required. The process was further complicated due to the numerous exceptions to the basic general rules and due to possible interactions between the windfall profit tax rules, the personal and corporate income tax rules, energy regulations, and state and local tax and energy laws.

The windfall profit tax also appeared to be a significant administrative burden for the IRS. The tax statute itself encompassed 13 sections in 25 pages of the 1986 Internal Revenue Code.[43] In addition, the IRS had to promulgate dozens of separate regulations, revenue rulings, letter rulings, and information releases to enforce it. Furthermore, there had been statutory amendments to the WPT in virtually every tax bill enacted between 1980 and 1988.

The IRS acknowledged the administrative burden of the tax in 1981 hearings before the House Subcommittee on Government Operations. A 1984 GAO report seemed to support this when it referred to the tax as "perhaps the largest and most complex tax ever levied on a U.S. industry."[44] Fortune magazine referred to the tax as one of "the most monumental excises ever levied in U.S. history... ."[45]

Reasons for Repeal of the Windfall Profit Tax

As was discussed in the background section, the crude oil windfall profit tax was repealed in August of 1988 — two and one-half years before the legislated termination date in January 1991. There was no one reason for repeal of the tax. Rather, repeal was caused by the confluence of several factors and conditions from 1987-1988. Yet, the anti-WPT sentiment was fairly widespread and numerous bills were introduce to repeal the tax.[46]

The Congress became convinced that the tax was a complex and costly tax to comply with and to administer. It was a compliance burden to the oil-producing industry and an administrative burden for the Internal Revenue Service even though, after FY1986, the tax generated little or no tax revenues. It is doubtful that the Congress would have repealed the WPT had it been generating significant revenues at that time or had it been expected to generate significant revenue in the future. The fact that the tax was generating little or no revenue, however, made the argument that the tax was a burden easier to accept.

Another apparent reason for the repeal of the WPT was the recognition that the tax kept domestic oil production below what it would have been without the WPT and increased petroleum imports above the level of imports without the WPT.[47] This made the United States more dependent upon foreign oil and therefore more vulnerable to either a price upsurge or a supply disruption. Petroleum imports were growing. From 1985 to 1986, there was a sharp increase in the share of oil use being met by imports. Oil imports as a percent of total U.S. oil consumption increased from 32% to nearly 38% — one of the largest annual increases on record. Moreover, projections showed this degree of dependence rising to over 50% by 1990, a projection which has been realized.[48] Today, petroleum imports account for nearly 60% of consumption and are projected by the Department of Energy to rise to nearly 70% by 2025.

Finally, the domestic U.S. oil industry was experiencing difficult economic conditions due to the collapse of oil prices in 1986. Crude oil prices dropped from about $30 per barrel in the fall of 1985 to just over $10 per barrel in the summer of 1986. After 1986, oil prices were volatile but basically increasing. At the time of repeal, oil prices were about $18 per barrel. Prices increased during the Persian Gulf war of 1992 but collapsed again in the fall/winter of 1998/1999.

There was little question about the effect of the rapid price decline on the U. S. oil-producing industry. It had a strong negative effect on oil producers (i.e., drillers, operators, and landowners with an economic interest in oil) in general, and the small independent producer in particular. According to industry data, earnings from exploration/production operations of selected companies in the first half of 1986 declined by about 60% from the first half of 1985.[49] This decline mirrored, roughly, the percentage decline in crude oil prices. After that, profits started to recover, especially for independents.[50]

Declining profits from oil production sharply reduced drilling and exploration expenditures and employment. In the long run, oil production was expected to decline significantly. Two states in particular, Texas and Louisiana, were hit hard by low crude oil prices. In these states, oil and oil dependent businesses (such as banks and other financial institutions) became bankrupt, large numbers of employees were laid off, and revenues to State and local governments plummeted. According to the Bureau of Labor Statistics, the oil and gas extraction industry nationwide lost about 130,000 jobs from the second quarter of 1985 to the second quarter of 1986.[51] Between 1982 and 1988, this industry lost about one-third of its jobs.[52] However, the collapse of oil prices helped some segments of the industry such as independent refiners and marketers.

The Congress came to view the windfall profit tax as a burden on an industry that was becoming severely depressed due to the sharp drop in oil prices and due to the volatility in oil prices. Repealing the WPT did not reduce industry tax payments so it was of little actual economic benefit at that time because oil prices were below base prices and there was no tax liability to producers. However, higher oil prices in the future might have exceed base prices and the WPT would have been triggered. At the very least, repealing the WPT reduced business costs and improved industry profitability somewhat by eliminating the compliance burden of the tax.

Opponents of repeal basically made the following arguments: (1) the oil industry's income is an economic rent or monopoly profit to a highly concentrated industry which society, through taxation, should share in; (2) the oil industry benefits from other tax subsidies which have traditionally kept effective income tax rates very low; (3) if oil prices rise above base price levels, then the tax would generate additional revenues which are badly needed to reduce large federal budget deficits; (4) the administrative apparatus is already in place and it makes little sense to eliminate the tax now, given that the tax is temporary. This final argument in favor of retaining the WPT was, in effect, a counterargument to those who have criticized the tax as a compliance and administrative burden. The point was that, even admitting its complexity, the WPT system was already in place. Much of the costs of administering the tax were fixed costs — they had been, in large part, already incurred, since most of the regulations had been promulgated. Given that the IRS had already incurred the fixed costs of running the WPT system, and given that the system would only be in effect for seven more years, they argued it made little sense to eliminate it.

WINDFALL PROFIT TAX LEGISLATION IN THE 109TH CONGRESS

There are currently 14 bills in the 109th Congress to impose some type of WPT. These fall generally into one of two categories: those that impose an excise type of WPT and those that would impose an income type. In addition, the bills differ in the way the WPT receipts would be used. Some bills would allocate the receipts to offset the cost of supplemental spending bills targeted to aid victims of Hurricanes Katrina and Rita. Others would allocate them to the highway trust fund to compensate for any losses from the proposed commensurate reduction in motor fuels excise taxes to offset the WPT. Several bills would appropriate the proceeds for the Low-Income Home Energy Assistance Program, which gives grants to poorer households to offset high energy bills and for residential weatherization.

Excise Tax Type of WPT

The excise tax type of WPT would generally impose an excise tax equal to 50% of the windfall profits not reinvested in either (1) oil/gas exploration and drilling, (2) refineries, (3) renewable electricity property, or (4) facilities for producing alcohol fuels or bio-diesel. Windfall profit would be defined as the difference between the market price of oil (at the wellhead) and a base price of $40/barrel, which would be adjusted for inflation. The bills that would impose this type of tax are S. 1631, H.R. 3752, H.R. 4203, H.R. 4248, H.R. 4449, H.R. 4263, S. 1981, and S. 2103. S. 1631 (Dorgan) was offered as an amendment to S. 2020, the Senate's version of tax reconciliation now in conference, but it was ruled out of order.

Three of the WPT bills are variations on this type of tax. For example, H.R. 2070 (Kucinich), H.R. 3664 (Kanjorski), and H.R. 3544 (DeFazio) would impose a graduated excise tax with the rates — 50%, 75%, or 100% — dependent on the extent to which profits exceed a reasonable level, as determined by a specially created board or commission. Also, the tax would be imposed on the windfall profit from and natural gas (and products thereof). These bills differ, however, on how the tax's proceeds would be used.[53]

Income Tax Type of WPT

The income tax type of WPT would impose a 50% tax on the excess of the adjusted taxable income of the applicable taxpayer for the taxable year over the average taxable income during the 2000-2004 period. The 50% tax would apply to crude producers and integrated oil companies with sales in 2005 or 2006 above $100 million. The tax would be temporary and apply to petroleum products as well as crude oil. The two bills that would take this approach are S. 1809 (Schumer) and H.R. 4276 (Larson) in the House. Senators Schumer and Reed sponsored S. 1809 as an amendment to S. 2020 (S.Amdt. 2635, and S.Amdt. 2626). In both cases, the amendments were ruled out of order.

A variant of the income tax type of WPT is H.R. 3712 (McDermott). This bill would impose a 100% tax on any profit above a 15% rate of return from the sale of crude oil, natural gas, or products of crude oil and natural gas. Revenues would be used to fund a program of gas stamps, which would be similar to the current federal food stamp program;

Economic and Policy Issues

This is not the first time policymakers have proposed a WPT on crude oil. In just two months from the beginning of July 1990 to August 1990, domestic oil prices (the spot price of West Texas Intermediate) nearly doubled increasing from just over $16 per barrel to nearly $32 per barrel. This was sufficient to prompt policymakers to call for reinstatement of the 1980 windfall profit tax.

These bills raise a number of economic and policy issues. The remaining sections discuss some of the more important economic issues surrounding proposed legislation, and draw relevant policy implications. The final section discusses alternative policy options.

Current Market Conditions

The first issue is the difference between the current market conditions and those when the 1980 WPT was imposed. As noted above, the 1980 WPT was imposed as part of compromise to decontrol crude oil prices — a *quid-pro-quo*. From a control regime level of about $6/barrel, crude prices were allowed to rise gradually to market levels (as influenced strongly by OPEC), which at that time were about $24/barrel. By contrast, today there are no price controls on crude oil and prices are determined in a generally competitive market, one in which the United States is a price taker, and one in which OPEC plays a relatively smaller (but still important) role. Crude oil prices have increased for significantly different reasons than was the case in the 1970s.

Also, on August 8, 2005, President Bush signed a comprehensive energy policy bill (P.L. 109-58) that provides over $2.5 billion in energy tax breaks (over 11 years) to the domestic oil and gas industry. Some have questioned the wisdom of providing additional tax subsidies to the oil industry at a time of high crude and petroleum product prices, and have proposed either reducing or eliminating these subsidies in lieu of the WPT.

The Question of Windfall Profits

Of course the fact that current high prices are due to reasons different than in the 1970s does not belie that there may be windfalls. Crude prices marched steadily upward during 2004 (averaging $37/barrel) and reached new highs in 2005.[54] The price of West Texas Intermediate (WTI) reached $55/barrel in October 2004, a 60% increase over the January 2004 price of $34/barrel. Prices reached $60/barrel during the summer of 2005 and peaked at nearly $70/barrel in August 2005.[55] After a short stay in the high $50's, prices hovered in the high $60's for several months, and in early 2006 stood at about $65/barrel.[56] Petroleum product prices have also spiked and continue to remain high. Increases in crude prices, in effect increase the value of oil reserves, and increase the revenue from the sale of domestically produced oil just as any higher market price for a commodity makes inventories or other stock of that product more valuable.[57]

The extent to which recent high petroleum prices have increased profits has been reported in the press and is discussed in two recent CRS reports; it is not discussed here.[58] A more intractable issue is whether any increased recent profits are true windfalls, and whether windfalls accrue to both crude oil and refinery products. The sharp and rapid increases in prices, if they are realized as higher profits, would be a pure windfall in the sense that they are an unforeseeable, unanticipated gain that accrues to owners of the Nation's stock of oil reserves. Most of any additional revenue and profit would accrue to the major oil companies since they own most of the reserves and produce the bulk of the oil in the United States. In a sense they are unearned: little or no additional cost or effort is incurred in generating this additional income; oil that would have been produced at $26 or $37 per barrel, can now be sold for $62 per barrel or more. In another sense they are earned: increased profits are the reward for the risks the industry takes to provide petroleum products to consumers.

As an illustration, using domestic oil production of 5.5 million barrels per day, a doubling of crude oil prices from $35/barrel to $70/barrel would generate an additional $70 billion in annual revenue to oil producers. If the baseline started at $10/barrel and prices rose to $70/barrel and stabilized at that rate for one year, revenues would increase by about $120 billion. Profits would also increase, but not by the increased revenue because income taxes — federal, state, and local — would have to be paid on the windfall. Thus, even without a windfall profit tax government tax revenues would increase commensurate with any oil industry windfall. This is because as profits of the oil producers increase so does taxable business income. Since there is little or no cost incurred in generating the added profits, then all of the revenue gains likely would be taxable. At a marginal tax rate of 35% — the marginal corporate tax rate — the federal government would gain about 35% of any oil industry windfall.[59]

In addition to the higher profits accruing to oil producers, there appear to be additional profits accruing to refiners, but these increases have been smaller than the profits accruing to producers. But these profits are not true windfalls as they are in the case of crude oil. While refinery margins and profits can and have increased, even sharply, they are due to the constraints on the supply capacity — essentially the supply or cost curve — of the refinery industry.

Another problem is one of timing. There is no question that producers and refiners could reap enormous windfall profits if the recent price upsurge is sustained. But it is not a certainty that the recent price spike will be sustained. Recently crude oil prices, while still high, declined somewhat from their peak of nearly $70 per barrel. Even though they are high there is no guarantee that they will not fall, possibly precipitously, and become losses in the near future if crude and product prices decline. And barring any other problems abroad, or natural disasters, prices would be expected to decline somewhat further. It should be remembered that oil prices declined from a high of over $30 per barrel in the early 1980s to about $10 per barrel at their low point in the spring of 1986. From 1986-1999 oil prices averaged about $17.00 per barrel, but fluctuated between $12 and $20 per barrel — the fluctuations including the most recent four-year period have been even greater. Domestic crude oil prices reached a low of about $8/barrel in December 1998, among the lowest crude oil prices in history after correcting for inflation. From this volatile behavior of oil prices derives the volatility of oil industry profits, and it is not clear that there will be a persistent windfall profit over the longer run.

Economic Implications

Depending on how the tax were structured, reinstating the windfall profit tax might make the United States more dependent upon foreign oil. This is likely to be a more serious problem today than in 1980 because, unlike then when the

United States was importing about 40% of its petroleum use, the nation is now importing close to 60%.

As the above analysis suggests, a windfall profit tax in the form of an excise tax — H.R. 3752, and S. 1631 — could reduce domestic oil production and increase the demand for imported oil and petroleum products. In economic terms, oil producers would view the tax as an increase in the marginal cost of domestic oil production. During the 1980s, the marginal cost (the incremental cost) of producing every barrel of taxable crude oil was higher with the WPT than without it. In consequence, the WPT reduced the supply of domestic oil to some extent; at every possible market oil price, it is estimated that domestic oil production was lower with the WPT than it would have been without it. The tax increased the marginal cost of producing domestic oil, thereby reducing domestic production and increasing the demand for oil imports both in the short run and long run. This is a result of the fact that oil imports to the United States are a residual, the difference between aggregate demand for oil and aggregate domestic oil supply. Any condition or factor which either reduces domestic supply (such as higher industry excise taxes) or which increases the aggregate demand for oil (such as higher national income) will increase oil imports. Crude oil prices might increase somewhat as the demand for imported oil increases, because the export supply of oil to the United States is somewhat elastic. The size of the effect would depend on the magnitude of the excise tax and the price elasticities of oil demand and supply in the United States, as well as the oil export supply to the United States.

Reinstating the windfall profit tax would probably raise the question of the burden and cost of compliance and administration of the tax, which was an important rationale for repeal of the tax in 1988. This issue seems to become more of a concern when WPT revenues don't live up to expectations. Unless the tax were greatly simplified, or unless a large amount of revenue were generated from the tax, past experience suggests that this could be a serious problem in reinstating the WPT. Many analysts believe, based on market fundamentals, that the recent oil price upsurge will not be a lasting one, and since oil prices have been highly volatile in recent years, there is no way to judge whether WPT revenues would be large enough to justify the alleged high costs of compliance and administration.

Alternative Policy Options

If, instead, an excise tax were to be imposed broadly on both imported as well as domestically produced oil (as proposed in the early 1980s by the Reagan Administration) much greater price effects would be expected — the price of crude oil in the United States would tend to be much higher than under the WPT on domestic oil alone. This is because the tax is imposed on imports, which are the marginal source of oil supplies and therefore the benchmark for crude oil prices. In this case, the tax would both increase the marginal costs of oil imports (an upward shift in the oil import supply schedule) and cause a slight movement along the schedule due to increased demand for oil imports, and thus have a bigger price effect.

An excise tax holiday — suspension of the 18.4¢/gallon tax on gasoline — combined with an equal revenue WPT on oil would be completely counterbalanced or offsetting. Eliminating the gasoline tax would cause refiners to reduce prices over time by the amount of the tax (or somewhat less depending on tax incidence, which depends on the ratio of price elasticities of the demand and supply schedules), but the WPT on all crude oil (which remember is actually an excise tax) would be shifted as a higher price of crude oil bought by refiners, thus offsetting the decline in product prices.[60]

From an economic perspective, the only tax that would be relatively neutral in the short run — that would have no (or few) price effects and other economic effects — would be a pure corporate profits tax, since this tax does not affect marginal production costs, and cannot be shifted in the short run. Thus, to the extent that a surtax on the corporate income of crude oil producers on their upstream operations could approximate such a tax, this would not raise crude oil prices and would not increase petroleum imports *in the short run.* While the current corporate income tax is not a pure corporate profits tax, a surtax for oil companies would arguably be an administratively simple and economically effective way to capture estimated oil windfalls in the short run. In the long run however, all taxes distort resource allocation and even a corporate profit tax (either of the pure type or the surtax on the existing rates) would reduce the rate of return and reduce the flow of capital into the industry, adversely affecting domestic production and increasing imports.

TECHNICAL APPENDIX

The algebraic formula for the WPT liability is as follows:

$$T_t = \tau\,[P_t^m - P_t^b(1+\pi_{t-2})](1-s)(1-c)$$

Where

T_t = the WPT in \$/barrel, in time t (e.g., in the 1st quarter of any year),
T = the WPT tax rate in percent (see Table 2 in the text),
P_t^m = the market (removal) price of domestic crude oil at the wellhead at time t,
P_t^b = the base price corresponding to the type of oil produced, as specified by law (see Table 2)
π = rate of inflation, as measured by the GNP deflator, so that Bt-2 is the GNP deflator lagged two quarters,[61]
s = the rate of State severance tax, if any
c = the rate of federal corporate income tax

ENDNOTES

1 Wage and price controls were authorized under the Economic Stabilization Act of 1970. 2 The Yom Kippur was occurred in October 1973, and the Arab oil embargo lasted from October 1973 to April 1974. The Emergency Petroleum Allocation Act also created the Crude Oil Entitlement Program and other programs intended to subsidize and protect small refiners. For a discussion of that program see CRS Report RL32248, *Petroleum Refining: Economic Performance and Challenges for the Future*, by Robert Pirog.

3 For more detail on the oil price control program see Kraft, John and Mark Rodekohr. *Crude Oil Price Controls: Their Purpose and Impact.* The Denver Journal of International Law and Policy, winter 1979. pp. 3 15-333.

4 Hakken, John. Excess Profits Tax. *The Encyclopedia of Tax Policy*. Joseph J. Cordes, and Jane Gravelle, eds. The Urban Institute Press, 1999. pp. 108-111.

5 U.S. Congress. Senate. *Energy Tax Provisions, 4: Crude Oil Equalization Tax and Rebate*. Committee Print. Senate Finance Committee. September 19, 1977. U.S. Govt. Print. Off. Washington, 1977.

6 Executive Office of the President. Office of Management and Budget. *Budget of the United States Government: Fiscal Year 1988*. January 1987. U.S. Govt. Print. Off. Washington. p. 2-42; and Executive Office of the President. Office of Management and Budget. *Budget Document of the United States Government: Fiscal Year 1989*. February 1988. U.S. Govt. Print. Off. Washington. p. 4-16.

7 The tax was repealed by § 1941(a) of P.L. 100-418, The Omnibus Trade and Competitiveness Act of 1988.

[8] P.L. 96-223 also contained amendments to the energy tax credits, as well as to several non-energy tax provisions. See Joint Committee on Taxation. *General Explanation of the Crude Oil Windfall Profit Tax Act of 1980 (H.R. 3919, 96th Congress; Public Law 96-223).* Joint Committee Print. U.S. Govt. Print. Off. Washington, 1981.

[9] An independent producer was one that was not a integrated producer, i.e., one with no more than 50,000 barrels per day in refinery capacity and $5 million or less in revenue from any gasoline retailing operations.

[10] This was oil from *any* well subject to price controls the revenues of which were used to finance enhanced oil recovery projects up to $15 million. That is, only oil that generated $15 million in revenues or less for such projects was exempt from the WPT, and only through September 31, 1981. Tertiary oil is oil that is recovered through certain enhanced oil recovery techniques such as flooding the reservoir with hot water or steam, or gases such as carbon dioxide. This is oil that is usually not recoverable through secondary recovery methods.

[11] CRS Report 88-147, *Oil Price Projections and the Windfall Profit Tax on Crude Oil*, by Salvatore Lazzari.

[12] U.S. Congress. House. *Crude Oil Windfall Profit Tax Act of 1979.* Report of the Committee on Ways and Means. Report No. 96-304, June 22, 1979. U.S. Govt. Print. Off. Washington, 1979.

[13] Pacific Northwest Laboratories. *An Analysis of Federal Incentives Used to Stimulate Energy Production.* Prepared for the U.S. Department of Energy. December 1978. P.226.

[14] According to the Joint Committee on Taxation, repealing the two oil and gas tax subsidies would increase tax revenues by about $1 billion per year. See CRS Issue Brief IB 10054, *Energy Tax Policy*, by Salvatore Lazzari.

[15] For a brief discussion of the energy tax provisions in H.R. 6, see CRS Issue Brief IB 10054, *Energy Tax Policy*, by Salvatore Lazzari.

[16] A few representative studies include Harberger, Arnold C. *The Taxation of Mineral Industries*. In U.S. Congress. Joint Committee on the Economic Report. *Federal Tax Policy for Economic Growth and Stability.* Joint Committee Print, 84th Congress, 1st session. November 9, 1955. Washington, U.S. Govt. Print. Off., pp. 439-449. Steiner, Peter O. *Percentage Depletion and Resource Allocation*. In U.S. Congress. House. Committee on Ways and Means. *Tax Revision Compendium*. Committee Print, 86th Congress, 1st session, vol. 2, November 16, 1959. Washington, U.S. Govt Print. Off., p. 949.

[17] Much of this early empirical evidence is cited in U.S. Congress. Senate. Committee on Interior and Insular Affairs. *An Analysis of the Federal Tax Treatment of Oil and Gas and Some Policy Alternatives*. Committee Print, 93rd Congress, 2nd session. Washington, U.S. Govt. Print. Off., 1974. p. 18.

[18] CRS Report 77-23 8, *Tax Provisions and Effective Tax Rates in the Oil and Gas Industry*, by Jane Gravelle.

[19] *Effective Federal Tax Rates on Income from Oil and Gas Extraction.* Typed Report by Jane Gravelle, April 13, 1983.

[20] Gravelle, Jane G. *Effective Federal Tax Rates on Income from Oil and Gas Extraction.* Paper presented at the annual meeting of the Conference for Taxation, Resources and Economic Development. October 1983. Cambridge, Mass. p. 6.

[21] Rates for integrated oil companies ranged from 6% to 15%; rates for independent producers ranged from 5% to 14%. This includes the effect of the minimum tax, which basically raises the rate, and repeal of the windfall profit tax, which basically lowers the rate. See Lucke, Robert and Eric Toder. *Assessing the U.S. Federal Tax Burden on Oil and Gas Extraction.* Energy Journal, vol. 8, no. 4, 1987. CRS calculations showed an effective marginal tax rate of 17% for integrated oil and gas producers. The rate for independents was not reported. See also U.S. General Accounting Office. *Tax Policy: Additional Petroleum Production Tax Incentives Are of Questionable Merit.* GAO/GGD-90-75, July 1990. Washington, 1990. p. 58.

Statutory marginal tax rates varied, and generally declined, during the life of the windfall profit tax from 46% in 1980 to 34% in 1988, when the WPT expired.

[22] *Corporate Income Taxes in the 1990s*. Institute on Taxation and Economic Policy, October 2000.

[23] *Corporate Income Taxes in the Bush Years*. Institute on Taxation and Economic Policy, September 2004.

[24] Gruber, Jonathan and Joshua Rauh. *How Elastic is the Corporate Income Tax Base?* National Bureau of Economic Research. June 2005, Table 1.

[25] Executive Office of the President. Office of Management and Budget. *Budget of the U.S. Government, FY2006: Historical Tables.* Table 1.2, p. 24.

[26] The Economic Recovery Tax Act of 1981 (P.L. 97-34) made several changes to the windfall profit tax which reduced revenues. The Tax Equity and Fiscal Responsibility Act of 1982 (P.L. 97-248) increased the tax on Alaskan oil which increased revenues by about $150 million per year. The Technical Corrections Act of 1982 (P.L. 97-448) made very minor changes in the windfall profit tax which reduced revenues negligibly.

[27] Berry, John M. And Art Pine. *Conferees Approve $227.7 Billion Oil Tax.* Washington Post, February 2, 1980. p. A1. Under the original House and Senate Finance Committee bills government would have received 75% and 54% of revenues from decontrol and a WPT, respectively (25% and 46% for industry, respectively). See U.S. Congressional Budget Office. *The Windfall Profits Tax: A Comparative Analysis of Two Bill.* Staff Working Paper. November 1979. p. xviii.

[28] Bureau of National Affairs. *Daily Tax Report.* JCT Staff Memorandum to Members of Senate Finance and House Ways and Means Committees on Windfall Profits Tax Revenue Estimates. March 23, 1981. Washington. p. J-1.

[29] From 1991-1998, the WPT revenues were a negative $1. 146 billion, representing IRS refunds and adjustments due to overpayment and over-withholding on prior returns, i.e., prior production. WPT revenues during this period were also affected by IRS efforts and legal cases to claim back taxes from companies that allegedly underestimated the market oil prices and consequently underpaid the Treasury. See Rose, Frederick. "ARCO Says IRS Asks $1 Billion for Back Taxes." *The Wall Street Journal.* July 19, 1988. p. 6. Thus, the actual gross revenues from 1990-98 were $78.923 million ($80.07 billion - $ 1.146).

[30] The need for revenue became even greater in the early 1980s than in the middle 1970s as budget deficits began to mushroom. As a result of the 1981-82 recession, tight monetary policy, the large tax cuts in 1981, and continued spending increases, the federal budget deficits were extremely large — over $1 trillion cumulatively for the period FY198 1-FY1 986. Between FY1986 and FY1987, the annual budget deficit dropped from $221.2 billion to $150.4 billion, but later it was projected to increase again. Deficits were large relative to our overall economy — in FY1985 the deficit as a share of GNP was about 5%, higher than any time in the 1970s.

[31] It is important to underscore the point that net revenues are not reported on any tax return — they must be estimated from tax return data on gross revenues and marginal personal and corporate tax rates. The WPT statute required the Treasury Department to estimate net revenues. These are the figures shown in column (4) of Table 4.

[32] There may be some small price effects if the export supply curve is not perfectly elastic. [33] This is discussed in detail in two other CRS reports: CRS Report 86-637, *Energy Taxes: A Comparative Analysis of the Gasoline Excise Tax and an Oil Import Tax and Their Effect on the States*, by Salvatore Lazzari; and CRS Report 86-572, *Oil Import Taxes: Revenue and Economic Effects*, by Bernard A. Gelb and Salvatore Lazzari.

[34] This estimate is quite sensitive to the assumed supply price elasticity, which is also unknown and has been derived from other studies. Generally, the more price elastic is the supply of oil, the larger would be the additional oil imports induced by the WPT.

[35] Dahl and Duggan (1996) survey many, but not all elasticity studies. See Dahl, Carol and Thomas Duggan. *U.S. Energy Product Supply Elasticities: A Survey and Application to the U.S. Oil Market.* Resource and Energy Economics, vol. 18, October 1996. pp. 243-254.

[36] For example, Mancke (1970) estimates a supply price elasticity ranging between 1 and 2. See Mancke, R. B. The Long Run Supply Curve of Crude Oil Produced in the United States. *Antitrust Bulletin.* Winter 1970. Kaplan, Seymour. *Energy Economics: Quantitative Methods for Energy and Environmental Decisions.* McGraw-Hill, New York. 1983. p. 67. More recent econometric studies tend to show less elastic supply curves. For example, Hogan (1989) estimates a long run supply curve of 0.58. See Hogan, William W. *World Oil Price Projections: A Sensitivity Analysis.* Energy and Environmental Policy Center. John F. Kennedy School of Government. Harvard University, 1989.

[37] Applied Management Sciences, Inc. *The Nonconventional Liquid Fuels R&D Analysis System: A Microcomputer-Based World Oil Market Model.* February 1, 1985. Prepared for the Department of Energy.

[38] These incentives were more than offset by other factors, including environmental regulations, low rates of return, and other factors inhibiting the flow of capital into refining and marketing. For a discussion see CRS Report RL32248, *Petroleum Refining: Economic Performance and Challenges for the Future*, by Robert Pirog.

[39] In a several reports, the GAO stated that the WPT was a complex tax. For example, in a 1984 report the GAO states: "The tax is very complex in design and operation and requires interaction among producers, operators, and withholding agents." See U.S. General Accounting Office. *Response to Questions About the Windfall Profit Tax on Alaskan North Slope Oil.* GAO /GGD-85-12. December 10,1984, p. 1. See also: U.S. General Accounting Office. *IRS's Administration of the Crude Oil Windfall Profit Tax Act of 1980.* GAO/GGD84-15, June 18, 198; and U.S. General Accounting Office. *Uncertainties About the Definition and Scope of the Property Concept May Reduce Windfall Profit Tax Revenues.* GAO/GGD-82-48, May 13, 1982.

[40] U.S. General Accounting Office. *IRS's Administration of the Crude Oil Windfall Profit Tax of 1980.* Report to the Chairman, Subcommittee on Commerce, Consumer, and Monetary Affairs, House Committee on

Government Operations. GAO/GGD-84-15, June 18, 1984. Washington, 1984. p. i.

[41] U.S. General Accounting Office. *Uncertainties about the Definition and Scope of the Property Concept May Reduce Windfall Profit Tax Revenues*. Report to the Secretary of the Treasury. May 13, 1982. GAO/GGD-82-48. Washington, p. 14.

[42] The net income limitation restricted the windfall profit tax liability to no more than 90% of the net income per barrel of oil. Net income was defined in terms of taxable income per barrel, with some adjustments.

[43] U.S. Code Title 26, Internal Revenue Code Sections 4986-4990.

[44] U.S. General Accounting Office. *IRS's Administration of the Crude Oil Windfall Profit Tax*. p. 1.

[45] Chapman, Stephen. Government's Windfall from Windfall Profits. *Fortune*, March 24, 1980. p. 60.

[46] There were 18 such bills in the 100th Congress.

[47] See, for example, U.S. Congress. Senate. *Energy Taxation Issues.* Hearings before the Subcommittee on Energy and Agricultural Taxation of the Committee on Finance, 100th Congress. January 30, 1987.

[48] CRS Report 87-779, *Oil Import Taxes: An Economic Analyis of S.694, The Economic Security Act of 1987*, by Salvatore Lazzari.

[49] Beck, Robert J., and Glenda E. Smith, "Unparalleled Drop in Crude Prices Reduces Earnings for OGJ Group," *The Oil and Gas Journal*, vol. 84, no. 35, Sept. 1986, pp. 17-22.

[50] Beck, Robert J. "Without Texaco, OGJ Group Earnings Increase 25.1% in 1987." *Oil and Gas Journal*, vol. 86, March 28, 1988. p. 6.

[51] U.S. Department of Labor. Bureau of Labor Statistics. *Monthly Labor Review*, vol. 109, no. 8, August 1986. p. 6.

[52] U.S . Department of Labor. Bureau of Labor Statistics. *Monthly Labor Review*, vol. 111, no. 3, March 1988. p. 72.

[53] H.R. 3544 (DeFazio) would impose price controls on gasoline, ban drilling in the Arctic National Wildlife Refuge, mandate minimum levels of inventory of crude oil and petroleum products, ban the export of Alaskan oil, and facilitate the draw down of the Strategic Petroleum Reserve. H.R. 2070 (Kucinich) would fund income tax credits for the purchases of fuel-efficient passenger vehicles, and to allow grants for mass transit.

[54] As noted the new highs are when measured in nominal terms (i.e., in current dollars). In real terms, the all time high was above $90/barrel reached in April 1980

[55] In August 2005 prices on the futures market, for the near term contract, reached $69.81/barrel, the highest price ever recorded on the Nymex exchange.

[56] September 2005 also witnessed the biggest one-day surge in oil prices on record: $4.39/barrel.

[57] An example is the stock of homes. As the price of new homes increases — due to inflation, increase in relative prices, or other reasons — the value of the existing stock of homes also increases so that homeowners experience a windfall.

[58] See CRS Report RL33021, *Oil Industry Profits: Analysis of Recent Performance*, by Robert L. Pirog.

[59] The marginal statutory rate is the appropriate rate to use in this instance rather than the marginal effective tax rate. The marginal rate includes the effects of intangible drilling costs and other tax provisions which enter into the calculation of the tax on income from the marginal *investment*. In the windfall profits tax situation in the text, there is no marginal investment, hence these oil and gas tax incentives/subsidies do not enter into the calculation.

[60] Eliminating the gasoline tax would deny the Highway Trust Fund of its principal source of revenue unless some adjustment were made. See CRS Report RL30497, *Suspending the Gas Tax: Analysis of S. 2285*, by Salvatore Lazzari.

[61] Except that for Tier III oil, the base price increases by 2% per year, so that the quarterly inflation adjustment for Tier III base prices is Bt-2 + .005.

In: The Oil Industry and Windfall Profits
Editor: Maria B. Coswell
ISBN 978-1-60692-840-0

Chapter 2

OIL INDUSTRY FINANCIAL PERFORMANCE AND THE WINDFALL PROFITS TAX*

Salvatore Lazzari and Robert Pirog

SUMMARY

Over the past 10 years, surging crude oil and petroleum product prices have increased oil and gas industry revenues and generated record profits, particularly for the top five major integrated companies (also known as the "super-majors"): Exxon- Mobil, Royal Dutch Shell, BP, Chevron, and Conoco/Phillips. These companies, which reported a predominant share of those profits, generated more than $100 billion in profits on nearly $1.5 trillion of revenues in 2007. From 2003 to 2007, revenues increased by 51%; net income (profits) increased by 85%. Oil output by the five majors over this time period declined by more than 2%, from 9.85 to 9.63 million barrels per day. Being largely price-driven, with no increase in output, and with little new production resulting from increased oil industry investment, many believe that a portion of the increased oil industry income over this period represents a windfall and unearned gain, i.e., income not earned by any additional effort on the part of the firms, but due primarily to record crude oil prices, which are set in the world oil marketplace.

* This is an edited, excerpted and augmented edition of a CRS Report for Congress publication, Order Code RL34689, dated September 30, 2008.

Numerous bills have been introduced in the Congress over this period to impose a windfall profits tax (WPT) on oil. Most of the bills were introduced in the 109th and 1 10th Congresses, after the enactment of the Energy Policy Act of 2005, which provided oil and gas industry tax incentives, in addition to the industry's traditional tax subsidies. An excise-tax based WPT would tax only domestic production, and like the one in effect from 1980-1988, would increase marginal oil production costs, which theoretically could reduce domestic oil supply, and raise petroleum imports, making the United States more dependent on foreign oil, undermining goals of energy independence and energy security. By contrast, an income-tax based WPT would be more economically neutral (less distortionary) in the short-run: sizeable revenues could be raised without reducing domestic oil supplies. Neither the excise- tax based or income-tax based WPT are expected to have significant price effects: neither tax would increase the price of crude oil, which means that refined petroleum product prices, such as pump prices, would likely not tend to increase.

In lieu of these two types of WPT, an administratively simple way of increasing the tax burden on the oil industry, and therefore recouping some of any excess or windfall profits, particularly from major integrated producers, would raise the corporate tax rate by, for instance, repealing or reducing the domestic manufacturing activities deduction under IRC § 199. This deduction is presently 6% of a firm's net income) and is available generally to all domestic manufacturing businesses (service firms are excluded), including almost all oil firms. Repealing this deduction for the major integrated oil companies, and freezing it at 6% for the remaining qualifying oil companies is estimated by the Joint Committee on Taxation to generate about $10 billion over 10 years.

INTRODUCTION

Over the past 10 years, the price of crude oil has been increasing, volatile, and has recently attained record high levels. The results of those price increases, which led to high gasoline prices, have been a weakening of the U.S. economy, and financial hardship for many American families who have also been buffeted by the housing slowdown, the credit crunch, rising unemployment, and other economic factors. While much of the American economy has suffered as a result of high oil prices, those prices generated record profit levels for the oil industry. Five companies, Exxon-Mobil, Royal Dutch Shell, BP, Chevron, and ConocoPhillips earned a predominant share of those profits

Record oil and gas industry profits have raised the concern of many public policy experts and federal policymakers, including many in Congress, who have questioned whether these profits were justified, or whether they constituted a "windfall" to the industry: an excessive, unearned, and unfair gain. Important factors in considering this issue might include the ultimate source, or reason, for the price increases, and what was the industry's role in generating them, i.e., whether it was through the direct result of the industry's efforts, in terms of employing its resources, decision-making, or risk-taking, or whether it was the result of fortuitous factors and events. Also important to the public policy question might be the actual size of the profits and what the industry did with them. If an industry invests profits into increased production capacity, the increased supply may ultimately cause prices to fall and the profits to dissipate. As the analysis in this report shows, the experience between 2003 and the first half of 2008 is not encouraging. Investments in oil exploration and development have not managed to keep company output from declining by some 7% over the five-year period.

Numerous bills have been introduced in the Congress over this period to tax the oil and gas industry's record profits. These bills generally take one or more of four approaches. First, some bills have proposed rescinding, or taking back, the tax incentives or subsidies enacted under the Energy Policy Act of 2005 (EPACT 05, P.L. 108-58), which not only expanded preexisting oil and gas industry tax subsidies, but enacted several new ones.[1] To be sure, EPACT05 also reinstated two excise taxes on petroleum that were larger than the tax subsidies,[2] but policymakers asked: 1) Why should the federal government provide *any* subsidies at a time when oil and gas prices and oil and gas industry profits are not only rising, but reaching higher and higher record levels; and 2) Why should the level of oil and gas industry subsidies be increased for an already profitable industry at a time when consumers and energy intensive industries (trucking, airlines, etc.) and the general economy are burdened by the high crude oil and gasoline prices?

Another policy option proposed in Congress to raise the tax burden on the oil and gas industry would repeal or cutback the pre-existing oil and gas industry tax subsidies, i.e., those that predated EPACT05, some of which were also expanded under EPACT05, and which generally are still in effect. Since the inception of the federal income tax system in the early 20th Century, the oil and gas industry has benefitted from sizeable — some estimates exceed $100 billion in real terms — federal tax subsidies, primarily for upstream activities such as exploration and development, and extraction and production, as compared to the downstream activities such as refining and marketing. For example, the domestic oil and gas industry is able to expense, i.e., write off in the first year (as compared to capitalize), intangible drilling costs (IDCs), and independent producers qualify for the percentage depletion allowance, rather than cost depletion. And while these traditional subsidies have been significantly pared back over the years, and are not now large relative to the size of the industry, some feel that no subsidies should be available to the industry at a time of surging record profits.

A third policy option that would raise taxes on oil and gas is to amend or reform certain income tax code provisions that, although legally and economically not subsidies — they would not be defined as subsidies in the tax expenditure sense by the Joint Committee on Taxation — may confer undue or disproportionate (and unnecessary) tax benefits to the oil and gas industry.[3] As an example, some bills propose to repeal the industry's use of the "last-in/first-out" (LIFO) system of inventory accounting under IRC § 472. This method values the goods sold as the most recent inventory purchase. During a period of rising prices, this method of inventory accounting increases production costs and reduces taxable income and tax liabilities.[4]

A second example is the proposed reform of the foreign tax credit. Many of the major energy tax bills that the Congress has considered over the last two years proposed to reform the tax treatment of foreign oil and gas extracting income (FOGEI) and foreign oil-related income (FORI). The proposed reforms are apparently focused on reducing instances in which such income is not measured correctly (overstated), so as to overstate foreign income taxes, and overstate their foreign tax credits, which tends to reduce the U.S. tax liability (i.e., the tax on U.S. source income).

Finally, a fourth option — the one that is discussed in this report — is to impose a windfall or "excess" profits tax i.e., a supplemental or additional tax on the oil industry, one based on windfall or excess profits in addition to other income or other taxes that the industry might pay. While some congressional Legislators envision this as a totally new type of tax on windfall gains, others would model it after the windfall profits tax on oil that existed from 1980 to 1988, while still others propose a higher corporate income tax burden, either through raising the corporate rate or eliminating deductions. Either way, a windfall profit tax would be in addition to the current tax on corporate and business income that applies to the oil and gas industry, and which tax profits at rates as high as 35%.[5]

This report discusses the fourth option: the windfall profit tax. The first section analyzes the major oil companies profit performance, particularly from 2003 to 2008, both in terms of earnings and how those earnings have been used. The second section is a brief legislative history of windfall profit tax proposals and legislation in both the 109th and 1 10th Congresses. The third section analyzes the idea of a windfall profits tax, including experiences with the tax of the 1 980s, its viability, and potential role in the tax system and economy.

OIL INDUSTRY FINANCIAL PERFORMANCE

During the Fall/Winter of 1998/1999, the spot market price of West Texas Intermediate (WTI) crude oil hovered between $11 and $12/barrel.[6] In July 2003, it was $30.75 per barrel. Five years later, in July 2008, the spot market price of WTI was $133.37, an increase of 334% in five years, and more than 1,000% over the 1998 trough. In June 2008, the spot market price reached the all-time high of $ 147/barrel.[7] This nearly 10-year period of generally increasing oil prices began after a poor year for profitability in the oil industry, 1998, and included another poor year, 2002. The increase in prices that began in late 2003 seemed to be largely unanticipated by the industry. It has since been attributed primarily to 1) a growing world economy, particularly the ascending economies of China and India, and 2) declining excess production capacity, particularly within OPEC (Organization of Petroleum Exporting Countries) producers. As the price of oil rose, company revenues, net incomes, and income taxes paid also rose, with Exxon-Mobil eventually becoming the most profitable corporation in the history of American industry.

Table 1. Financial Data for Oil Industry Firms, 2003-2008 (billions of dollars)

Income Statement Item	2003	2004	2005	2006	2007	2008 First Half
Revenue	1,144.6	1,396.4	1,620.2	1,718.9	1,868.4	1,253.9
Net Income	72.4	100.7	139.8	162.8	155.8	87.6

Source: Profit Profile Supplements, Oil Daily.

Note: The companies included in Table 1 are ExxonMobil, Royal Dutch Shell, BP, Chevron, ConocoPhillips, Marathon, Hess, Occidental, Murphy, Encana, Apache, Devon, Andarko, XTO, EOG, Noble, Chesapeake, Pioneer, Newfield, Valero, Sunoco, Tesoro, Western, Frontier, Holly, and Alon and the companies they merged with, or acquired, since 2003.

The oil industry is composed of thousands of companies, ranging from the major integrated oil companies with operations around the globe, independent producers (which can be very large), to relatively small oil service and equipment companies. The rise in oil prices over the past 10 years — particularly over the past five years — has enhanced the profitability of virtually all sectors of the industry, directly, or indirectly.[8]

Table 1 reports three measures of the financial performance of the domestic oil and gas industry in the United States, using data for the largest oil and gas producing companies as reported by Oil Daily. Note that while industry revenue (price times output), increased by 63%, from $1.1 trillion to nearly $1.9 trillion from 2003 to 2007, industry profits (net income) more than doubled, increasing from $72 billion to more than $150 billion over this time period.

The Super-Major Integrated Oil Companies

While the oil and gas industry's good fortune has been extensive and widespread, it has also been concentrated among industry's largest firms. Table 2 uses more recent financial data to show that the performance of the industry is dominated by the five largest firms (the "super-majors"): ExxonMobil, Royal Dutch

Shell, BP, Chevron, and ConocoPhillips.[9] For the first half of 2008, the net income of these five firms constituted 90% of the total net income of the nine firms considered to be integrated oil companies operating in the United States. Of the set of twenty-six firms, which includes integrated oil companies, independent oil and gas producers, and independent refiners and marketers, the five major firm's net income was 82% of the total. As a result of the dominant position of these five firms in the industry, the data in the remainder of this report are limited to their financial performance.

The data in Table 2 show that the revenue of the five major firms increased. This increase was caused primarily by the increasing prices of oil and petroleum products. Total revenue is measured as price times the quantity of goods and services sold. In the case of the five major oil companies over this period, the increase in revenues was largely price-driven, with quantities produced largely stagnant. For example, in 2003, ExxonMobil produced 2.59 million barrels per day (b/d) of crude oil, and in 2007, ExxonMobil produced 2.61 million b/d, an increase of less than 1/10th of 1%. In general, the five major oil firms were unable to produce more crude oil and petroleum products in response to the higher prices as shown in Table 3. Note that only one company, BP, produced more crude over this period, and that, for the five firms as a whole, output declined by 2.2% from 2003-2007, and by 5.8% through the first half of 2008.

The revenue data for the first half of 2008 reflects the sharp increases in the price of oil observed during the first six months of the year, with the price of WTI reaching $147 per barrel in June. During the third quarter of 2008, the price of oil declined from the June peak, making extrapolation of the first half 2008 to the entire year uncertain. The likely result is that revenue increases were predominately price- driven, which, as noted, is important in considering a windfall profit tax.

Table 2. Revenue of the Top Five Major Integrated Oil Companies, 2003-2008 (billions of dollars)

Company	2003	2004	2005	2006	2007	2008 First Half
ExxonMobil	246.7	298.0	371.0	377.6	404.5	254.9
Shell	269.1	265.2	306.7	318.8	355.8	245.7
BP	236.0	294.8	253.6	270.6	291.4	201.1
Chevron	120.0	155.3	198.2	210.1	220.9	148.9
ConocoPhillips	105.0	136.9	183.4	188.5	194.5	129.9
Total	**976.8**	**1,150.2**	**1,312.9**	**1,365.6**	**1,467.1**	**980.5**

Source: Profit Profile Supplements, Oil Daily.

Table 3. Crude Oil Production by the Major Oil Companies (million barrels per day)

Company	2003	2004	2005	2006	2007	2008 First Half
ExxonMobil	2.51	2.57	2.52	2.68	2.61	2.43
Shell	2.39	2.25	2.09	2.03	1.82	1.73
BP	2.12	2.53	2.56	2.47	2.41	2.43
Chevron	1.81	1.71	1.67	1.73	1.76	1.65
ConocoPhillips	1.02	0.90	0.91	1.13	1.03	0.94
Total	**9.85**	**9.96**	**9.75**	**10.04**	**9.63**	**9.18**

Source: Profit Profile Supplements, Oil Daily.

Table 4 shows the net income of the five major oil companies over the period 2003 to 2008. While revenues increased by 51% from 2003 to 2007, net income increased by 85%. These percentages suggest that price increases for crude oil and petroleum products increased at a higher rate than costs and taxes for the five major firms. Analogously to the observed increase in revenues, the increase in net income experienced by the five firms was largely price-driven.

Table 4. Net Income of the Major Oil Companies, 2003-2008[a] (billions of dollars)

Company	2003	2004	2005	2006	2007	2008 First Half
ExxonMobil	21.5	25.3	36.1	39.5	40.6	22.6
Shell	12.7	18.5	22.9	25.4	27.6	15.7
BP	16.4	16.2	19.3	22.2	17.3	13.4
Chevron	7.2	13.3	14.1	17.1	18.7	11.1
ConocoPhillips	4.7	8.1	13.5	15.5	11.9	9.6
Total	**62.5**	**81.4**	**105.9**	**119.7**	**116.1**	**72.4**

Source: Profit Profile Supplements, Oil Daily.

a. Data reflect consolidated worldwide earnings of these firms. Data segmenting net income on the basis of geographical earnings are not available.

Table 5. Average Profit Rates In the Oil Industry, 2003-2008 (percent)

	2003	2004	2005	2006	2007	2008 First Half
Major Integrated Companies	6.4	7.0	8.0	8.7	7.9	7.4
All Oil Industry	6.3	7.2	8.6	9.5	8.4	7.0

Source: Profit Profile Supplements, Oil Daily. CRS calculations.

The previous measures of financial performance focus on absolute levels: absolute levels of revenue, and absolute net income. Perhaps a better measure of performance is the profit rate, which may also be measured in variety of ways. One measure of the profit rate is the operating margin, essentially net income divided by revenue. As a result of the increasing price of oil driving up both total revenues and net incomes, the return on revenue, or the profit rate, increased slightly for both the major integrated oil companies as well as for the industry as a whole, as shown in Table 5. This indicator of industry performance is not out of line with the rate in the manufacturing industry generally, which in 2007 has a profit rate of 8.9%. However, a more appropriate measure of the relative profit rate may be the rate of return on equity (ROE). Using this measure of profit rate, the oil and gas industry's ROE was, on average, significantly greater than the ROE for the manufacturing industry generally. According to the Energy Information Administration (EIA), the oil and gas industry earned a 27% ROE in 2006, down slightly from 2005, but more than 9% higher than the average ROE for all manufacturing companies.[10]

For the entire 2000-2006 period, the oil and gas industry's ROE averaged 7 percentage points higher than manufacturing's ROE, while for the 1985-1999 period, the oil and gas industry's ROE was only 2 percentage point higher. By this measure, the industry's recent high profits, measured both in absolute terms, and relative to ROE, suggest the presence of excess or windfall profits.

The data in Table 1, Table 2, Table 3, and Table 4 characterize an industry that was unable to respond to the market signal of higher price by increasing output as predicted by economic theory. As a result, their revenues and net incomes increased proportionately, possibly supporting the judgement that those profits constituted a "windfall," at least in the sense that they were not earned through output expansion or improvement, risk-taking, or investments leading to cost reductions in production. However, the data in Table 5 could be taken to suggest that the major oil companies did not earn more net income respective to the value of their product, than many other industries, and the value of their product was determined on a world market, beyond their control. From this point of view oil was attaining its fair market value.

The net income data presented in Table 4 was earned in three sectors of the industry: upstream operations (the exploration and production of oil and natural gas), downstream operations (the refining, transportation, distribution, and marketing of petroleum products, including motor gasoline, diesel, jet fuel, and other petroleum products), and chemicals and all other "non-oil" activities. Over time, the relative importance of these sectors may shift in terms of potential to generate net income. For example, while downstream activities were very strong in 2005 and 2006, they weakened in 2007 and 2008. In terms of general corporate income taxation, this cyclical pattern of change is likely to have little effect, since it is over-all corporate net revenues that form the tax base. In terms of possible windfall profit taxation, however, this cyclical pattern of change might be important as it is net-income from crude oil ownership and production that is likely to have a windfall gain, or "unearned" income component.

Use of Profits[11]

In economic theory, when firms earn returns in excess of the market rate of return, they are likely to re-invest in their businesses to expand output and improve their technologies to meet the challenge of new firms that might choose to enter the industry as new competitors. In the theory of corporate finance, firms that seek to maximize shareholder value use profits to invest in business projects that offer a higher potential rate of return than that currently earned by the firm.

Increased Investment

Capital expenditures for the five major oil companies were $48.6 billion in 2003, $48.7 billion in 2004, $57.2 billion in 2005, and more than $80 billion in 2006 and 2007, for a total increase of 77% over the five-year period. This increase is proportionately less than the increase in net income over the period. Part of total investment funding is directed to environmental compliance for both facilities and products, and does not increase capacity to bring petroleum products to the market. [12] It is well known that the five companies have not committed to the construction of a new refinery in the United States since the 1970s, though existing refineries have been expanded and upgraded. When the demand for gasoline exceeds the ability of the U.S. refineries to produce, the gap has been filled with imported product.

Increased Oil Output

Although the oil industry, and the five major firms have invested in exploration, development, and production, those investments have not met with noticeable success. Table 2 shows that the companies have failed to expand, or even maintain their oil production rates.

Cash Reserves and Dividend Payouts

The major oil companies accumulated cash reserves from 2003 through 2007. The five firms held $19.4 billion in cash in 2003 and $52.27 billion in 2007, an increase of almost 170%. While cash holdings declined in 2007, compared to 2006, this was the result of reduced balances at only one company, Chevron.[13] The other four major oil companies continued to build their cash balances.

From one point of view, cash balance accumulation gives firms flexibility and positions them to take advantage of opportunities quickly. It is also likely that the rapid increase in the price of oil and profits from 2003 to 2007 exceeded corporate plans and strategies on how to use it. However, the theory of corporate finance suggests that extraordinary cash returns to shareholders are appropriate only when the management feels that individual shareholders are likely to have access to higher return investment alternatives than management can identify. Activities that represented direct returns to stockholders, in the form of increased dividend payments and stock buy-backs also claimed a share of earned profits.

LEGISLATIVE HISTORY OF WINDFALL PROFITS TAX PROPOSALS

Almost from the outset of the surge in crude oil and petroleum product prices, many in Congress became concerned over the level of oil industry profits, and bills were introduced to raise taxes on the oil and gas industry by imposing some type of windfall profits tax. These proposals are viewed as a way of generating tax revenue to fund subsidies for low income persons and offset the burden of recent high petroleum prices, and programs for energy conservation and alternative and renewable fuels. Also, they are viewed from the perspective of basic equity or fairness — the overall distribution of the tax burden could be viewed as being more fair when taxes on the record, unearned incomes of business enterprises are used to reduce the tax burden of lower income persons burdened by the higher prices that contributed to those record earnings.

The concept of a windfall profits tax is not new; a tax on windfall, or excess, business profits has been one the instruments of fiscal policy, used by both state and federal governments, whenever business profits either rise too fast or rise to levels that are either considered too high, above "normal" or fair, or which reflect "excessive" rates of return. At the federal level, however, such taxes have been used sparingly — being viewed as extraordinary measures, their use limited to wartime or other periods characterized by economic emergencies and instabilities such as hyper- inflations. Such was the case with the surtax on business profits imposed as a temporary measure to control large profits earned during World Wars I and II, and the Korean War.[14]

A type of windfall profits tax on domestic crude oil production was in effect from April 1980 to August 1988. This tax, which was actually an excise tax, not a profits or income tax, was part of a compromise between the Carter Administration and the Congress over the decontrol of crude oil prices. It is discussed and analyzed in detail in CRS Report RL33305.[15] Some have proposed reinstating this tax, although it should be underscored that the current situation giving rise to possible windfall profits — the current reasons for the high price of petroleum products and record profits — is different than the conditions and rationale which existed at the time that tax was imposed.[16]

Reinstating the oil windfall profits tax was again discussed in 1990, when crude oil prices doubled in just two months due to the crisis in the Middle East (Iraq invaded Kuwait on August 2, 1990).[17] More recently, a windfall profits tax has also been discussed as part of the presidential campaign, with Senator Obama reportedly being a proponent of such a tax (in addition to cutting the industry's subsidies and providing a consumer energy credit to compensate for rising prices), and Senator McCain not discussing this option.[18]

Windfall Profits Tax Legislation in the 109th Congress

After the enactment of EPACT05 in August 2005, congressional interest in a windfall or excess profits tax on the oil and gas industry intensified. In fact, most of the bills to impose some type of windfall profits tax were introduced in the 109th Congress — there were more than a dozen such bills — after EPACT05 was enacted into law. Many of these bills proposed to use the revenues from the WPT to offset the burden of higher gasoline prices for consumers.[19] There were two types of windfall profits tax bills in the 109th Congress: those that would have imposed an excise tax on windfall profits based on the price of crude oil, and those that would have imposed an income tax on windfall profits based on either the existing tax law's definition of corporate taxable income or excessive rates of return.

Excise Tax Type of WPT

As noted above, the WPT that was in effect from 1980-1988 was not an income tax but an excise tax — it was not a type of tax that most economists would consider a true tax in "windfall gains or income." The tax was imposed on the difference between the market price of oil, which was technically referred to as the removal price, and a statutory 1979 base price that was adjusted quarterly for inflation and state severance taxes. Almost every barrel of domestically produced crude oil — i.e., every barrel of domestic production that was not specifically tax-exempt — was subject to this excise tax.

The excise tax type of WPT was the type proposed in most of the WPT bills in the 109th Congress. These bills would have generally imposed an excise tax equal to 50% of the windfall profits not reinvested in 1) oil/gas exploration and drilling, 2) increased refinery capacity, 3) renewable electricity property, or 4) facilities for producing alcohol fuels or bio-diesel. These bills would have defined a windfall profit as the difference between the market price of oil (at the wellhead) and an inflation-adjusted base price of $40/barrel — in other words, they would have effectively defined a windfall as the difference between the market oil price and $40. S. 1631, H.R. 3752, H.R. 4203, H.R. 4248, H.R. 4449, H.R. 4263, S. 1981, and S. 2103 were of this type. S. 1631 (Dorgan) was offered as an amendment to S. 2020, the Senate's version of tax reconciliation which went to conference, but was ruled out of order.

Income Tax Type of WPT

Some of the WPT proposals in the 109th Congress were of the income tax type, using the existing corporate income tax system to assess the tax, or defining the tax base in terms of taxable income under the existing corporate income tax.

Typical of the income tax type of WPT were those that would have imposed a 50% tax on the excess of the adjusted taxable income for a taxable year over the average taxable income during the 2000-2004 period. The 50% tax would have applied to crude producers and integrated oil companies with sales in 2005 or 2006 above $100 million. The tax would have been temporary and would apply to petroleum products as well as crude oil. S. 1809 (Schumer) and H.R. 4276 (Larson) in the House were of this type. Senators Schumer and Reed sponsored S. 1809 as an amendment to S. 2020 (S.Amdt. 2635 and S.Amdt. 2626). In both cases, the amendments were ruled out of order.

A variant of the income tax type of WPT is H.R. 3712 (McDermott). This bill would have taxed any profit from the sale of crude oil, natural gas, or products of crude oil and natural gas above a 15% rate of return at 100%. Tax revenues would have been earmarked for a program of gas stamps to help indigent persons offset the burden of recent high gasoline prices, which would be similar to the current federal food stamp program.

Other Types of WPT Proposals

Some WPT proposals are not easily classified. For example, H.R. 2070 (Kucinich), H.R. 3664 (Kanjorski), and H.R. 3544 (DeFazio) would have imposed a graduated tax with the rates — 50%, 75%, or 100% — dependent on the extent to which profits exceed a reasonable level, as determined by a specially created board or commission. These bills differ, however, on how the tax's proceeds would be used.[20] Whether these WPT would have been excise tax based or income tax based, or whether they would have used some other tax base, is unknown since the bills did not provide a definition of either profits or a reasonable profit.

Windfall Profit Tax Legislation in the 110th Congress

WPT proposals have been introduced in the 1 10th Congress, although there are fewer excise tax based and more income tax based than in the 109th Congress. Also, many of the income tax based proposals have focused on repealing the IRC § 199 deduction for domestic manufacturing activities, which is effectively equivalent to an increase in the existing marginal corporate tax rate — and is not really tax on windfall profits. As discussed in more detail below, the corporate income tax system could be used as an administratively simple way to increase the tax burden on the oil and gas industry, and approximate a WPT without the risks of adverse economic and energy market effects.

Excise Tax Type of WPT

S. 1238 (Casey) would impose a 50% excise tax on any major integrated oil company based on the difference between the market price of oil and $50 (this $50 would be adjusted annually for inflation). This bill is the only bill to propose an excise type of WPT in the 1 10th Congress.

Income Tax Type of WPT

Table 6. Tax Payments by the Major Oil Companies, 2005-2007 (billions of dollars)

	2005	**2006**	**2007**
ExxonMobil	23.30	27.90	29.86
Shell	17.99	18.31	18.65
BP	9.29	12.31	21.17
Chevron	11.10	14.84	13.48
ConocoPhillips	9.91	12.78	11.38
Total	**71.59**	**86.14**	**94.54**

Source: Company Income Statements, available at [http://www.hoovers.com].

On the Senate side two bills introduced by Senator Clinton, S. 701 and S. 2971, would tax half of the excess profits of major integrated oil producers, and oil producers with gross revenues above $100 million per year. Excess profit would be defined as taxable income above 110% of the average taxable incomes over the 2000-2004 period. Thus, these bills would use the existing tax law definition of taxable income, which, as discussed below, not only facilitates tax administration and compliance, but also minimizes economic distortions and adverse, short-run economic effects. S. 2991 and S. 3044 (Reid) take a similar approach to the Clinton bills, but would 1) tax excess profits at a lower rate, 25% instead of 50%, and 2) use 200 1-2005 (2002-2006 in S. 3044) as the base period rather than 2000-2004. Also, any increased investment in renewable energy over the same base period would be credited toward the tax and hence, reduce the windfall profit tax liability. S. 3044 (Reid) is the Democratic leadership bill; it was introduced on June 11, 2008.[21]

Other Types of WPT Proposals

H.R. 5800 (Kanjorski) and H.R. 6000 (Kucinich) would tax excess profits above a "reasonable" amount at rates ranging from 50% to 100%. A Reasonable Profits Board would be created to determine the reasonable profit level. The tax base to which this tax would apply appears to be the income tax, but it is not defined.

ANALYSIS OF ECONOMIC AND POLICY ISSUES

Even with rising crude oil and petroleum product prices, and record oil industry profits, not all of those profits or returns constitute a windfall or unearned income — much of that income may be a return to investment and its capital stock, a return on the firm's decision-making and risk-taking. And, indeed, the business income taxes paid to state and local governments, the federal government, and even foreign countries, which are substantial, reflects their share of these profits in the form of taxes. Table 6 shows the total taxes reported in their financial statements by the major integrated oil companies in 2005, 2006, and 2007. These include income, excise (including motor fuel excise taxes to federal, state, and local governments), and severance taxes (which are primarily state levies). The record profits earned by the five major oil companies generated liability for tax payments that increased by 32% from 2005 through 2007 as shown in Table 6. Any windfall profits tax that might be adopted would add to these tax revenues.[22]

Such taxes, however, do not belie the existence of windfall gains or unearned income, and do not necessarily undermine the case for a WPT. A well designed and structured WPT tax, however, would tax only the true windfall component of oil industry incomes. Moreover, such a tax would be simple to administer and comply with, and would avoid or minimize any adverse economic and energy market effects. As the discussion below suggests, while in theory the concept of a windfall profit seems simple and intuitive, in practice it can be difficult to measure accurately and so the actual implementation of a WPT involves a compromise over differing fiscal policy objectives: 1) administerability — collecting the excess revenues (or windfall gains) in the least costly manner in terms of tax administration and compliance, 2) economic efficiency — devising and structuring the tax in a way that minimizes economic distortions, including adverse output and price effects, and adverse impacts on petroleum imports, energy independence, and energy security.

The remaining sections of this report discuss some of the more important economic issues surrounding proposed legislation, and draw relevant policy implications. The final section discusses alternative policy options.

Defining and Measuring Windfall Gains

The most vexing problem in designing a WPT is how to define a true windfall, i.e., the tax base. In theory, there is a difference, even if a subtle difference, between a windfall gain, excess profit, and an unearned gain. A windfall gain applies to income or wealth that is unexpected, a gain which arrives fortuitously, possibly due to factors outside of the control of the institution in question. This can be distinguished from excess profits, which might be considered to be more subjective, and are based on defining an acceptable profit, and attributing everything above that level as excessive. This difference might be illustrated with a hypothetical example: a firm might be so efficient that its above-average profitability (or returns) might be considered excessive, but not a windfall. However, if its production waste, which formerly was costly to dispose of, became valuable, that gain might be considered a windfall.

Finally, one might differentiate both a windfall profit and an excess profit from an economic rent, a technical term for the return to a fixed factor of production or resource. For such resources that are in fixed supply, economic analysis suggests that earned profits are not necessary for the services of that factor to be available to the market.

The record profits reported by the oil and gas industry have characteristics of both windfall gains, excess profits, and even economic rents, although they are more in the nature of "quasi-rents" than pure economic rents because some industry investment, effort, and risk is required for production: oil supply is not perfectly inelastic. Thus, for purposes of taxing this income — from the perspective of merely raising revenues — it matters little whether the tax is an excise tax or income tax type, or whether it is based on industry rates of returns compared to rates of return in other industries (manufacturing, for example), or in the general economy. Since the profits are price-driven, and since there is little or no output or investment effect, either the excise tax or income tax or rate of return type could be used to collect the windfall.

Rather, the critical difference in the type of WPT tax to implement is in the firm's economic response to the tax, and in the cost of administration and compliance costs. The next two sections of this report examine the output and price effects of the two different types of WPT. It also examines the effects on the demand for oil imports, and therefore energy independence, and energy security. Finally, this section examines a variety of other tax options for increasing the tax burden on the oil and gas industry, options that have been offered as alternatives to the WPT. In general the conclusions of this analysis are that an excise tax is non-neutral and could have adverse economic effects, particularly on the level of dependence on foreign oil, while the income tax approach is relatively neutral in the *short run*, thus minimizing economic distortions and other adverse economic effects.

The Non-neutral Economic Effects of the Excise Tax Type of WPT

As discussed in more detail below, an excise tax on domestically produced crude oil may only approximate a windfall profit tax, and may have certain adverse energy market and economic effects. In other words, depending on how a WPT were structured, an excise tax type of WPT — for instance, by reinstating the WPT of the 1980s — might make the United States more dependent upon foreign oil, which in turn might have implications for energy security.

Output Effects

Economic theory suggests that a WPT in the form of an excise tax — e.g., H.R. 3752 and S. 1631 in the 109th Congress; S. 1238 in the 1 10th Congress — could reduce domestic oil production.

In economic terms, oil producers would likely view the tax as an increase in the marginal, or incremental, cost of domestic oil production — the marginal cost of producing every barrel of taxable crude oil would be higher by the amount of the excise tax. An increase in the marginal cost of production might be viewed as an incentive to produce less oil. However, this effect is likely to be mitigated in the U.S. oil market by other factors. The theoretical analysis assumes that the difference between price and marginal cost is relatively small, implying that the imposition of a tax would reduce profits to, or below, the competitive level. In the current oil market, oil prices are typically far above the marginal cost of production, implying that even after paying a tax, profitability could remain high, continuing to provide an incentive for production. The economic analysis also assumes that some cheaper alternative source of oil is available to substitute for taxed domestic production. Firms are unlikely to pay the market price for oil in the international oil market instead of using domestic production. If they did this, they would lose the difference between the price and cost of domestic oil, substituting foreign oil, for which price is effectively equivalent to marginal cost. Finally, marginal production adjustments are unlikely to be made to existing oil wells. Once an oil well goes into production, the maximum sustainable flow rate is likely to simultaneously be the maximum economic flow rate, allowing for oil field management to be based on physical factors related to output.

Oil Imports and Energy Independence

If the domestic supply of oil were reduced in response to an excise tax on domestically produced oil, the demand for imported oil and petroleum products would likely increase, unless some other policy would concomitantly reduce the demand for petroleum to offset a tax-induced reduced supply. This is because oil imports to the United States are a residual, adjusting to reflect the difference between aggregate demand for oil and aggregate domestic oil supply.

Price Effects

One of the concerns — and one of the arguments made by opponents of a WPT — is that such a tax would raise prices: a WPT on crude oil would raise the price of crude oil, which would then be passed on to consumers in the form of higher petroleum product prices — higher prices of gasoline, diesel, jet fuel, and other products. If true, then this might defeat one of the purposes of imposing such a tax: to relieve consumers (both personal and business consumers such as truck drivers) from the burden that recent high gasoline and diesel price have had on them.

As noted, an excise tax only on oil produced domestically in the United States would increase marginal or incremental production costs, and, in theory, a profit maximizing firm would respond to this type of tax by reducing output and attempting to raise prices to offset the higher marginal production costs. However, in the case of domestic crude oil, the higher marginal cost cannot be shifted forward as a higher oil price, because oil is priced in the international (world) oil market).[23] Oil producers would not be able to shift the tax forward as a higher oil selling price because the first purchaser (generally, the refiner) would merely substitute imported crude, which would be tax-exempt. Instead, this type of WPT would reduce the net selling price paid to producers. As noted earlier, the first purchaser would subtract the tax from the price paid to the producer (supplier) — the producer's net selling price of each barrel of oil would be less by the amount of the WPT.

The Neutrality of the Corporate Income Tax Type of WPT

Output Effects

From an economic perspective, the income tax based WPT — such as those that use the existing corporate tax system to define excess or windfall taxable income — would be relatively neutral in the short run — it would have no (or few) output, or price effects and other economic effects. The reason for this is that a firm maximizes profit at the point at which market prices are equal to marginal production costs, and neither are affected in the short run by an increase in the corporate tax burden — the profit maximizing level of output and price are unaffected by the tax. Thus, to the extent that a surtax on the corporate income of crude oil producers on their upstream operations could approximate such a tax, this would not raise crude oil prices and would not tend to increase petroleum imports *in the short run.*

In the long run, however, all taxes (or almost all taxes) distort resource allocation, and even a corporate profit tax (either of the pure type or the surtax on the existing rates) would raise average long-term production costs, reduce the rate of return and reduce the flow of capital into the industry relative to other industries, and move resources away from the corporate form of business organization. All these effects could adversely affect domestic production, possibly resulting in increased petroleum imports.

Oil Imports and Energy Independence

Because the income tax type of WPT does not create incentives to reduce domestic production in the short run, there is no increase in the demand for imports under such a tax in the short run, although it could if the tax were still in effect in the long run.

Price Effects

From an economic perspective, that is to say, in theory, increasing marginal tax rates on corporate income would have no (or few) price effects. The reason is as before: a firm maximizes profit at the point at which market prices are equal to marginal production costs, and neither are affected by an increase in marginal tax rates — the profit maximizing level of price is unaffected by such a tax.

Alternative Policy Options

As mentioned above, should Congress decide to tax oil industry surpluses, there would be several alternative policy options to increase taxes on the domestic oil and gas industry, including eliminating the industry's several targeted tax subsidies, reducing tax benefits provided through the general or non-targeted provisions of the tax code, or even eliminating the tax incentives enacted under EPACT05. These options are discussed in a separate CRS report.[24] Rather, the remaining sections of this report address alternatives that are really variations on the two WPT options discussed before.

Rescinding the § 199 Deduction

One option that has been proposed frequently in congressional bills is the repeal of, or reduction in, the IRC § 199 deduction for domestic manufacturing activities. This option does not actually attempt to measure or tax windfall profits, but uses the existing corporate income tax system to effectively increase the marginal corporate tax rate on domestic oil and gas producers, targeting primarily the major integrated oil companies.[25]

Enacted in 2004 as an export tax incentive, this provision allows a deduction, as a business expense, for a specified percentage of the qualified production activity's income (or profit) subject to a limit of 50% of the wages paid that are allocatable to the domestic production during the taxable year. The deduction was 3% of income for 2006, is currently 6%, and is scheduled to increase to 9% when fully phased in by 2010. For the domestic oil and gas industry, which qualifies for this deduction (i.e., it is not excluded from claiming it) the deduction applies to oil and gas or any primary product thereof, provided that such product was "manufactured, produced, or extracted in whole or in significant part in the United States." Note that extraction is considered to be manufacturing for purposes of this deduction, which means that domestic firms in the business of extracting oil and gas from underground reservoirs or deposits qualify for the deduction. This deduction was enacted under the American Jobs Creation Act of 2004 (P.L. 108-357, also known as the "JOBS" bill). It was originally a substitute for repeal of the export tax benefits under the extra-territorial income tax exclusion, which was ruled to be in violation of trade laws.[26]

Repealing, or cutting back this deduction would be effectively equivalent to an increase in the marginal income tax rate, and as noted earlier a change in the corporate tax burden in the short run is relatively neutral. For example, at the top marginal corporate tax rate of 35%, which typically applies to large corporations such as major oil and gas producers and refiners, the current deduction of 6% is equivalent to a marginal corporate income tax rate of 32.9% (35% x 0.94) rather than 35%.[27] The proposed elimination of this deduction would be, thus, equivalent to an increase in the marginal tax rate from 32.9% to 35% for those major oil companies to which this would apply. All other large corporations would continue to face a top marginal tax rate of 32.9%, with the exception of non-manufacturing enterprises (services, for example), which do not qualify for the § 199 deduction. Thus, as noted, this option does not attempt to measure and tax the oil industry's windfall profits; it is just a way of using the existing corporate income tax system to increase the tax burden on the oil industry, and recoup some of the windfall or excess profits in the form of corporate income taxes.[28]

As before, eliminating the deduction — that is to say, raising the corporate tax rate — would increase total (or average) business costs and therefore reduce profitability among the major oil and gas producers. As long as marginal production costs are unaffected, there would be no price effects *in the short run.* Similarly, the demand for imports is likely to remain the same in the short-run. Thus, this type of corporate income tax increase would arguably be an administratively simple and economically effective way to capture at least some of the oil industry's windfalls in the short run. However, at a current deduction of 6%, and a marginal corporate tax rate of 35%, only a small portion of the industry's likely windfalls would likely to be captured under this option.

The market price of crude oil and natural gas, or even of refined petroleum products, such as gasoline, would not be expected to increase very much, if at all, by such a change in the short run. In general, also, the income tax increases are not expected to have real output effects in the short run, although they could cause resources to flow to other industries in the long run as long as these other industries are allowed the manufacturing deduction, which is equivalent to a lower marginal tax rate.

An Income Type WPT Tax and § 199 Repeal

Some WPT proposals combine the WPT, with a repeal or reduction in the § 199 manufacturing activities deduction. This effectively uses two instruments to raise the corporate tax rate on domestic oil and gas producers, which, as noted before, would be relatively neutral in terms of its price, output, and import effects. One bill that takes this approach is S. 3044, which would rescind the IRC § 199 deduction for major integrated oil companies, and assess a 25% tax on the difference between profits in any one year and 110% of the average of profits over the 2002-2006 period.[29] This bill follows many of the earlier bills in that it attempts to recoup for the federal taxpayer some of the windfall or "excess" profits reported by the oil and gas industry as a result of unprecedented high petroleum prices.

A Tax on Imported and Domestically Produced Crude Oil

If, instead, an excise tax were to be imposed broadly on both imported as well as domestically produced oil (as proposed in the early 1980s by the Reagan Administration), this tax would produce upward price effects — the price of crude oil in the United States would tend to be higher than under the WPT on domestic oil alone. This is because the tax would be imposed on imports, which are the marginal source of oil supplies and therefore the benchmark for crude oil prices. The effect on domestic production and the level of imports — dependence on foreign imports — would depend on the size of the tax and the price elasticities of domestic supply and import demand.

An Excise WPT and Gas Tax Suspension

An excise tax holiday — suspension of the 18.4¢/gallon tax on gasoline — combined with an equal revenue WPT on oil would be completely counterbalanced or offsetting. Eliminating the gasoline tax might cause refiners to reduce prices over time by the amount of the tax, depending on the state of the energy market at the time of suspension, (or somewhat less depending on tax incidence, which depends on the ratio of price elasticities of the demand and supply schedules), but the excise-tax based WPT on all crude oil (i.e., one imposed on both imports and domestic production) would be shifted as a higher price of crude oil bought by refiners, thus offsetting the decline in product prices.[30]

Possible Revenue Effects

A WPT on crude oil could generate sizeable revenues depending on the tax rate and the tax base, i.e., whether it was excise tax based or income tax based, and depending on crude oil prices at the time the tax were imposed. This section of the report does not provide revenue estimates, because CRS neither has the data nor the estimating models to do so.

The excise tax based WPT probably has the greatest revenue potential merely because of the high recent level of crude oil prices, and because the tax base is not adjusted for operating costs (operating costs are not deducted from the tax base) as in the case of the income-tax based WPT. To illustrate, S. 1238 (Casey) would tax any major integrated producer on the difference between market price and $50/barrel at a 50% rate. If oil prices remain over $100 per barrel, this tax could generate substantial revenues.

The income tax type of WPT generally has a smaller revenue potential, again because, 1) taxable income is used as the tax base, which is generally much smaller than book income (which is the net income measure reported in Table 4). Also, the income tax based proposal typically uses the average of taxable income over a lagged five-year period. Thus, in a period of generally rising prices, the base tends to progressively decline.

Table 7. Estimated Revenue Effects of Repealing the § 199 Deduction for All Oil and Gas Industry, and for Major Integrated Oil and Gas Producers (billions of dollars)

2008	2009	2010	2011	2012	2013	2014	2015	2016	2017	2008-12	2008-17
.262	.605	.776	.950	1.022	1.098	1.18	1.269	1.364	1.466	3.615	9.992

Source: U.S. Congress. Joint Committee on Taxation. Estimated Budget Effects of the Revenue Provisions Contained in Titles I. And XV. Of H.R. 6, The Clean Renewable Energy and Conservation Tax Act of 2007, as Passed by the House of Representatives on December 6, 2007. JCX-1 12-07, December 12, 2007.

Finally, there are revenue estimates related to repeal of the IRC § 199 deduction. These have been estimated several time in recent years by the Joint Committee on Taxation for several different congressional proposals. The figures reported in Table 7 are the most recent available, and show that over 10 years, nearly $10 billion in additional revenue could be collected from repealing the § 199 deduction for major integrated oil producers. It should be noted that the figures in Table 7 would likely change if they were re-estimated using more current assumption about crude oil prices, the state of the oil industry, the condition of the macro-economy, and other economic variables.

ENDNOTES

[1] In the 110th Congress H.R. 86, H.R. 498, H.R. 1945, and S. 115 propose to rescind EPACT's oil and gas industry tax breaks.

[2] EPACT05 reduced taxes (provided additional tax subsidies) valued at $2.633 billion over 11 years. It also reinstated two excise taxes: the Oil Spill Liability

Trust fund tax, and the Leaking Underground Storage Tank Trust fund tax, both of which are imposed on oil refineries. According to the Joint Committee on Taxation, over 11 years, the value of the tax increase was $2,857 million, $224 million ($2,857-$2,633) more than the tax cuts Also, the Tax Increase Prevention and Reconciliation Bill of 2006 (P.L. 109-222), enacted on May 17, 2006, reduced one of the industry's tax subsidies, i.e., it increased taxes on the oil industry, by about $189 million.

[3] There is an important economic distinction between a subsidy and a tax benefit. As is discussed elsewhere in this report, business firms, including oil and gas companies, generally receive a variety of tax benefits that are not necessarily targeted subsidies (or tax expenditures) because they are available generally.

[4] See CRS Report RL3 3578 for more detail.

[5] As discussed below on p. 18, the marginal corporate income tax rate is presently less than 35% for most domestic manufacturing activities due to the 6% domestic manufacturing activities deduction, which typically applies to large corporations such as oil and gas producers and refiners, and which is equivalent to a marginal corporate income tax rate of 32.9% (35% x 0.94) rather than 35%. In addition, present federal tax law imposes a minimum tax on corporations (and individuals) to the extent that their minimum tax exceeds their regular tax liability.

[6] The low of $10.82/barrel was reached on December 10, 1998.

[7] The $147 is in nominal or current dollars; crude oil prices also reached a record high in real terms, or inflation adjusted terms.

[8] There are, of course, exceptions. Stone Energy Corporation, for example, lost money (they reported negative net income) during 2006. See Standard and Poor's. *Industry Surveys: Oil and Gas, Production and Marketing.* March 20, 2008. P. 48.

[9] ExxonMobil, Chevron, and ConocoPhillips are U.S. based firms. BP is a British firm, and Royal Dutch Shell is a Dutch and British firm, both with U.S. subsidiaries.

[10] The EIA reports these data in its annual reports for the Financial Reporting System (FRS), which are based on detailed financial and operating data and information submitted each year to the EIA on Form EIA-28, the Financial Reporting System (FRS). The FRS Companies derive the bulk of their revenues and income from petroleum operations, which include natural gas production. A majority of these companies are multinational, with 40% percent of the majors' net investment located abroad. EIA supplements the FRS data with additional information from company annual reports and press releases, disclosures to the U.S. Securities and Exchange Commission, news

reports and articles, and various complementary energy industry data sets. See Energy Information Administration. *Performance Profiles of Major Energy Producers (Issues December 2007, March 2006, February 2004, and January 2002).*

[11] This section of the report is based on CRS Report RL34044, *The Use of Profit by the Five Major Oil Companies,* by Robert Pirog. See CRS Report RL34044 for a more complete analysis of this topic as well as more complete data.

[12] The Energy Information Administration, *The Impact of Environmental Compliance Costs on U.S. Refining Profitability 1995-2001,* May 2003.

[13] Balance sheet data for the five major oil companies can be obtained at [http://www.hoovers.com].

[14] Hakken, John. Excess Profits Tax. *The Encyclopedia of Tax Policy.* Joseph J. Cordes, and Jane Gravelle, eds. The Urban Institute Press, 1999. pp. 108-111.

[15] U.S. Library of Congress. Congressional Research Service. *The Crude Oil Windfall Profits Tax of the 1980: Implications for Current Energy Policy.* CRS Report RL33305, by Salvatore Lazzari. March 9, 2006.

[16] As noted above, the 1980 WPT was imposed as part of compromise to decontrol crude oil prices — a *quid pro quo.* From a control regime level of about $6/barrel before the tax, crude prices were allowed to rise gradually to market levels (as influenced strongly by OPEC), which at that time were about $24/barrel. By contrast, today there are no price controls on crude oil and prices are determined in a generally competitive global crude oil market, one in which the United States is a price taker, and one in which OPEC plays a relatively smaller (but still important) role. Also, more recently crude oil prices have increased for significantly different reasons than was the case in the 1970s. Unlike the 1980s when crude oil prices declined sharply to pre-decontrol levels just after the WPT was imposed (and for most of the life of the tax), crude oil prices since the trough of 1998/1999 have increased fairly steadily and consistently and have surpassed the levels of 1982 in real terms.

[17] From the beginning of July 1990 to August 1990, domestic oil prices (the spot price of West Texas Intermediate) nearly doubled increasing from just over $16 per barrel to nearly $32 per barrel.

[18] The Wall Street Journal. *The Economy: Competing Visions for Fixing Today — and Tomorrow.* August 25, 2008. P. R5, R8. Obama also supports repealing the IRC § 199 manufacturing deduction, which is also allowed for oil and gas companies, raising nearly $10 billion in revenues over 10 years, and another provision to limit the ability of oil and gas companies to claim foreign tax credits for their overseas operations could raise another $4 billion. According

to the Bureau of National Affairs, Inc. *Daily Tax Report*. "Obama Says Tax Plan Offers More Tax Cuts." Pp. GG1, Friday, August 29, 2008.

[19] Some of the bills would have allocated the receipts to offset the cost of supplemental spending bills targeted to aid victims of Hurricanes Katrina and Rita. Others would allocate them to the highway trust fund to compensate for any losses from the proposed commensurate reduction in motor fuels excise taxes to offset the WPT. Several bills would have appropriated the proceeds for the Low-Income Home Energy Assistance Program, which gives grants to poorer households to offset high energy bills and for residential weatherization.

[20] H.R. 3544 (DeFazio) would impose price controls on gasoline, ban drilling in the Arctic National Wildlife Refuge, mandate minimum levels of inventory of crude oil and petroleum products, ban the export of Alaskan oil, and facilitate the draw down of the Strategic Petroleum Reserve. H.R. 2070 (Kucinich) would fund income tax credits for the purchases of fuel-efficient passenger vehicles, and to allow grants for mass transit.

[21] S. 3044 would also roll back $17 billion in existing oil and gas industry tax breaks over 10 years for the largest oil companies; revenues would be earmarked to expanding renewable energy development. In addition to the tax provisions, S. 3044 would prohibit, and provide penalties for, price gouging by the oil and gas industry, tighten regulation of speculators in offshore oil, and suspend filling of the Strategic Petroleum Reserve.

[22] According to the U.S. Department of the Interior, the domestic oil industry in the United States also pays billions in royalties, which are not taxes but factor payments, the return to landowners on their mineral assets. For FY2007, total federal royalties on oil and gas (including natural gas liquids) were $9.4 billion. See the Minerals Management Services website at [http://www.mrm.mms.gov].

[24] See CRS Report RL33763, *Oil and Gas Taxes and Subsidies: Current Status and Analysis*, by Salvatore Lazzari.

[25] Repealing this deduction has been part of almost every major energy tax bill in the Congress over the last two years. Section 199 repeal is also part of the so-called "Gang of 16" draft energy bill (no number yet) which would also open up much of the Outer Continental Shelf to oil and gas production, and restrict speculation in the oil commodities (futures) markets.

[26] CRS Report RL32652, *The 2004 Corporate Tax and FSC/ETI Bill: The American Jobs Creation Act of 2004*, by David L. Brumbaugh.

[27] Corporations are currently taxed at 15% of the first $50,000 of taxable income, 25% of the taxable income from $50,001 to $75,000, 34% of the taxable

income from $75,001 to $10 million, and 35% of taxable income above $10 million.

[28] The proposed repeal or cutback in the § 199 deduction could also be a way of funding other provisions in the bills.

[29] S. 3044 also restricts the ability of major oil and gas companies to claim tax credits for taxes and other payments to foreign governments against the U.S. tax on foreign source income, which also is a way of increasing the corporate tax burden.

[30] Eliminating the gasoline tax would deny the Highway Trust Fund of its principal source of revenue unless some adjustment were made. See CRS Report RL3 0497, *Suspending the Gas Tax: Analysis of S. 2285*, by Salvatore Lazzari. and CRS Report RL34475, *Transportation Fuel Taxes: Impacts of a Repeal or Moratorium,* by John W. Fischer and Robert Pirog.

In: The Oil Industry and Windfall Profits
Editor: Maria B. Coswell
ISBN 978-1-60692-840-0

Chapter 3

OIL AND GAS TAX SUBSIDIES: CURRENT STATUS AND ANALYSIS*

Salvatore Lazzari

SUMMARY

The CLEAN Energy Act of 2007 (H.R. 6) was introduced by the House Democratic leadership to revise certain tax and royalty policies for oil and natural gas and to use the resulting revenue to support a reserve for energy efficiency and renewable energy. Title I proposes to repeal certain oil and natural gas tax subsidies, and use the resulting revenue stream to support the reserve. The Congressional Budget Office (CBO) estimates that Title I would repeal about $7.7 billion in oil and gas tax subsidies over the 10-year period from 2008 through 2017. In House floor debate, opponents argued that the cut in oil and natural gas subsidies would dampen production, cause job losses, and lead to higher prices for gasoline and other fuels. Proponents counterargued that record profits show that the oil and natural gas subsidies were not needed. The bill passed the House on January 18 by a vote of 264-123. This report presents a detailed review of oil and gas tax subsidies, including those targeted for repeal by H.R. 6.

The Energy Policy Act of 2005 (EPACT05, P.L. 109-58) included several oil and gas tax incentives, providing about $2.6 billion of tax cuts for the oil and gas

* This is an edited, excerpted and augmented edition of a CRS Report for Congress publication, Order Code RL33763, dated February 27, 2007.

industry. In addition, EPACT05 provided for $2.9 billion of tax increases on the oil and gas industry, for a net tax increase on the industry of nearly $300 million over 11 years. Energy tax increases comprise the oil spill liability tax and the Leaking Underground Storage Tank financing rate, both of which are imposed on oil refineries. If these taxes are subtracted from the tax subsidies, the oil and gas refinery and distribution sector received a net tax increase of $1,356 million ($2,857 million minus $1,501 million).

EPACT05 was approved and signed into law at a time of very high petroleum and natural gas prices and record oil industry profits. The House approved the conference report on July 28, 2005, and the Senate on July 29, 2005, clearing it for the President's signature on August 8 (P.L. 109-58). However, the tax sections originated in the106th Congress, with its effort in 1999 to help the ailing domestic oil and gas producing industry, particularly small producers, deal with depressed oil prices. Subsequent price spikes prompted concern about insufficient domestic energy production capacity and supply. All the early bills appeared to be weighted more toward stimulating the supply of conventional fuels, including capital investment incentives to stimulate production and transportation of oil and gas.

In addition to the tax subsidies enacted under EPACT05, the U.S. oil and gas industry qualifies for several other targeted tax subsidies (FY2006 revenue loss estimates appear in parenthesis): (1) percentage depletion allowance ($1 billion); (2) expensing of intangible drilling costs for successful wells and non-geological and geophysical costs for dry holes, including the exemption from the passive loss limitation rules that apply to all other industries ($1.1 billion); (3) a tax credit for small refiners of low-sulfur diesel fuel that complies with Environmental Protection Agency (EPA) sulfur regulations ($ 50 million); (4) the enhanced oil recovery tax credit ($0); and (5) marginal oil and gas production tax credits ($0).

ACTION IN THE 110TH CONGRESS

The CLEAN Energy Act of 2007 (H.R. 6) was introduced by the House Democratic leadership to revise certain tax and royalty policies for oil and natural gas and to use the resulting revenue to support a reserve for energy efficiency and renewable energy. The bill is one of several introduced on behalf of the Democratic leadership in the House as part of its "100 hours" package of legislative initiatives conducted early in the 1 10th Congress.

Title I proposes to repeal certain oil and natural gas tax subsidies, and use the resulting revenue stream to support the reserve. According to the Congressional

Budget Office (CBO), the provisions in Title I would make about $7.7 billion available for the reserve over the 10-year period from 2008 through 2017.[1]

H.R. 6 came to the House floor for debate on January 18, 2007. In the floor debate, opponents argued that the reduction in oil and natural gas incentives would dampen production, cause job losses, and lead to higher prices for gasoline and other fuels. Opponents also complained that the proposal for the Reserve does not identify specific policies and programs that would receive funding. Proponents of the bill counterargued that record profits show that the oil and natural gas incentives were not needed. They also contended that the language that would create the Reserve would allow it to be used to support a variety of research and development (R&D), deployment, tax incentives, and other measures for renewables and energy efficiency, and that the specifics would evolve as legislative proposals come forth to draw resources from the Reserve. The bill passed the House on January 18 by a vote of 264-123.

BACKGROUND

The Energy Policy Act of 2005 (P.L. 109-58), enacted on August 8, 2005, expanded some of the existing tax subsidies for the oil and gas industry and created several new ones.[2] The oil and gas tax incentives in EPACT05 were added on top of several existing special tax subsidies for oil and gas. The industry also benefits from provisions of current tax law that are not strictly tax subsidies (or tax expenditures) but that nevertheless provide advantages for and reduce effective tax rates of the oil and gas industry.

The remainder of this report discusses these tax provisions in detail. The first section, below, discusses the origin and evolution of the oil and gas tax subsidies that were incorporated into the 2005 act. The second section summarizes each of the oil and gas tax subsidy provisions in the 2005 energy act and reports its corresponding revenue loss estimate. Section three describes other oil and gas tax subsidies, those that existed before EPACT05 and were generally not affected by it. The final section describes several tax provisions that benefit the oil and gas industry; these are not tax subsidies per se — they are not considered to be tax expenditures — but are deemed by some observers to confer excessive (or unfair) benefits for the industry.

POLICY CONTEXT AND ANALYSIS

Tax incentives for oil and gas supply have historically been an integral (if not the primary) component of the nation's energy policy. The domestic oil and gas industry was granted three tax code preferences, or subsidies: (1) expensing of intangible drilling costs (IDCs) and dry hole costs, introduced in 1916; (2) the percentage depletion allowance, first enacted in 1926 (coal was added in 1932); and (3) capital gains treatment of the sale of oil and gas properties.[3] These tax subsidies reduced marginal effective tax rates in the oil and gas industries, reduced production costs, and increased investments in locating reserves (increased exploration). They also led to more profitable production, some acceleration of oil and gas production, and more rapid depletion of energy resources than would otherwise occur. Partially in response to tax incentives, but also due to the low cost of discovering and developing the huge new resource base, there were discoveries during the 1930s of vast reserves in Texas, which led to a period of overproduction of oil and gas and concomitant declines in prices, which led to demand to prorationing under the Texas Railroad Commission.[4]

Beginning in the 1970s and through much of the 1990s, energy tax policy shifted away from fossil fuel supply and moved toward energy conservation through both energy efficiency and the development of alternative and renewable fuels. However, rising and repeated spikes in petroleum prices that began around 2000 and were repeated over the next six years (combined with high and spiking natural gas prices, an electricity crisis, and blackouts) caused policymakers to focus on increasing energy production and supply of many diverse energy sources, including oil and gas.

The tax incentives for the oil and gas industry in the EPACT05 originated in the106th Congress's effort in 1999 to help the ailing domestic oil and gas producing industry, particularly small producers, deal with depressed oil prices. This situation fostered proposals for economic relief through the tax code, particularly for small independent drillers and producers. Proposals focused mainly on production tax credits for marginal or stripper well oil,[5] but they also included carry-back provisions for net operating losses, and other fossil fuel supply provisions.[6] Subsequent comprehensive energy policy legislation, including H.R. 4 in the107th Congress, proposed an expanded list of oil and gas tax incentives. The energy tax breaks in this bill (the Securing America's Future Energy Act of 2001, as approved by the House on August 1, 2001) were larger in terms of tax revenue loss than any other comprehensive energy policy legislation proposed during this period. They also were larger than those proposed in

EPACT05: $33.5 billion of energy tax cuts, compared with the $14.5 billion loss eventually enacted under P.L. 109-58.

Interest in incentives and subsidies was boosted by the belief that much of the crisis was caused by insufficient domestic production capacity and supply. All the early bills appeared to be weighted more toward stimulating the supply of conventional fuels, including capital investment incentives to stimulate production and transportation of oil and gas. These proposals were further repackaged and expanded into the first broadly based energy bills and comprehensive energy policy legislation, such as H.R. 6 in the 109th, that evolved further and ultimately became EPACT05.[7] The House approved the conference report on July 28, 2005, and the Senate on July 29, 2005, clearing it for the President's signature on August 8 (P.L. 109-58).

The 2005 act became law at a time of very high prices for crude oil, petroleum products, and natural gas, and record oil and gas industry profits. This engendered the enmity of the general public and congressional proposals to (1) revoke the incentives enacted under the 2005 act; (2) repeal or pare back the historical, but extant, tax subsidies and other tax advantages; and (3) impose sizeable new taxes on the industry such as a windfall profit tax.[8]

Public and congressional outcry did lead to a paring back of one of the tax subsidies liberalized in the 2005 act: two-year amortization, rather than capitalization, of geological and geophysical (G&G) activity costs, including those associated with abandoned wells (dry holes).[9] This exploration subsidy was the largest upstream tax subsidy (as opposed to a "downstream" or a refinery subsidy), in terms of federal revenue loss, enacted under the 2005 act, although it was and still is a relatively small tax subsidy. The Tax Increase Prevention and Reconciliation Act (P.L. 109-222), signed into law in May 2006, reduced the value of the subsidy by raising the amortization period for major integrated oil companies from two years to five years, still faster than the capitalization treatment before the 2005 act, but slower than the treatment under that act. Independent (nonintegrated) oil companies may continue to amortize all G&G costs over two years.

This relatively minor cutback has not muted the calls for rolling back oil and gas tax subsidies, as petroleum prices (and industry profits) remain somewhat high, particularly those of the biggest oil and gas companies. On September 1, 2006, the House Democratic leadership reportedly sent a letter to the House Speaker proposing a rollback of all of the 2005 energy act tax subsidies.[10] On October 25, 2006, then- House Democratic Leader Nancy Pelosi, urged the Congress to repeal those tax breaks.

Many bills were introduced in the 109th Congress to pare back or repeal the oil and gas industry tax subsidies and other loopholes. Many of the bills focused on the oil and gas exploration and development (E&D) subsidy — expensing of intangible drilling costs (IDCs). This subsidy, which has been in existence since the early days of the income tax, is available to integrated and independent oil and gas companies, both large and small alike.[11] It is an exploration and development incentive, which allows the immediate tax write-off of what economically are capital costs, that is, the costs of creating a capital asset (the oil and gas well). On September 18, 2006, Senators Wyden and Bennett introduced a bill (S. 3908) to give consumers a discount on the purchase of more fuel efficient vehicles that would have been paid for by reducing the IDCs deduction for major integrated oil companies. Comprehensive energy legislation (S. 2829) unveiled by Senate Democrats on May 17, 2006, would have not only eliminated expensing of IDCs, but would have also reduced several other tax benefits (or loopholes) to the oil and gas industry (such the foreign tax credits). The latter are not subsidies (or tax expenditures) in the strict sense of special tax measures unavailable generally, but as discussed below, some consider these unnecessary tax benefits nonetheless.[12] H.R. 5234 focused on repealing three of the seven fossil fuel tax provisions in the 2005 act: temporary expensing of equipment costs for crude oil refining, the small refiner exception to percentage depletion, and the amortization of geological and geophysical (G&G) costs. H.R. 5218 would have denied oil and gas companies the new domestic manufacturing deduction under IRC § 199.

There is speculation that in the 1 10th Congress, the Democratic leadership in both the House and Senate will begin to examine these breaks more closely, particularly because many of their legislative priorities (such as cutting back the increasingly heavy burden of the alternative minimum tax) will have to be paid for.[13]

OIL AND GAS TAX PROVISIONS IN EPACT05 AND THEIR REVENUE EFFECTS

EPACT05 included a plethora of spending, tax, and deregulatory incentives to stimulate the production of conventional and unconventional oil and natural gas, such as gas from Alaska, deep water oil and gas in the outer continental shelf, and oil from marginal wells or private and federal lands. These incentives include tax breaks, royalty relief, streamlined permitting procedures, and other measures. The tax incentives include approximately $14.5 billion over 11 years of incentives to both stimulate domestic production and distribution of fossil fuels and reduce the demand for these fuels through energy efficiency and production of alternative and renewable fuels.

Title XIII, subtitle B, of EPACT05 includes the tax incentives for fossil fuel supply — for production, transportation, and distribution — of oil and gas, as well as capital incentives for expanded refinery capacity. The subtitle does not include coal supply incentives, which are subsumed in the electricity infrastructure subtitle. Although many of the oil and gas tax incentives in EPACT05 are production tax credits and other such "upstream" production incentives, some are capital incentives for natural gas infrastructure (accelerated depreciation of natural gas pipelines). In total, the tax incentives alone are worth about $2.6 billion over 11 years to the industry (an average of about $250 million a year in tax breaks).[14]

Subtitle B, thus, applies specifically to the oil and gas industry, including the refinery industry, for increased supply incentives. Tax incentives are provided — again mostly by liberalization of existing tax code provisions. The incentives are both production incentives (i.e., tax benefits are based on quantities of oil and gas) and capital incentives (i.e., tax benefits are based on magnitude of capital investment, such as pipelines). Both unconventional and conventional oil and gas supply are targeted for tax cuts.

Amortization of Geological and Geophysical Expenditures

Firms engaged in the exploration and development (E&D) of oil and gas incur a variety of costs prior to actual extraction. The tax treatment of these "upstream" E&D costs differs depending on the specific type of activity and depending on whether they are incurred by an integrated or nonintegrated (i.e., independent) producer. An independent producer is defined by Internal Revenue Code (IRC) § 613A(d), as described below.

E&D costs may be generally categorized as four types. First, there are the geological and geophysical costs (G&G). These are exploratory costs (such as for seismic surveys) associated with determining the precise location and potential size of a mineral deposit. A second type of cost is the mineral acquisition or lease rights expenses — the costs of buying or leasing the land under which deposits are thought to exist — such as lease bonuses.

If a property is considered prospective for containing economically recoverable deposits of oil or gas, the firm drills exploratory (and, if successful, subsequently development) wells to ascertain the magnitude of the deposits. These activities have associated various types of drilling costs. Tangible drilling costs, the third type of E&D costs, are amounts paid for tangible drilling and nondrilling equipment such as drilling rigs, casings, valves, pipelines, and other tangible machinery and equipment that have a salvage value. Finally, there are intangible drilling costs, or IDCs as they are frequently called. IDCs are amounts paid by the lease operator for fuel, labor, repairs to drilling equipment, materials, hauling, and supplies. They are expenditures incident to and necessary for the drilling of wells and preparing a site for production of oil and gas. For example, roads may have to be constructed to move in derricks and other types of drilling equipment; often a camp may have to be built with residences to house employees. The power for the equipment and the water supplies are also IDCs. IDCs also may include the cost to operators of any exploratory drilling or development work done by contractors under any form of contract, including a turnkey contract.

In general, as noted above, prior to EPACT05, all four types of costs — G&G costs, mineral rights, tangible equipment, and intangible drilling costs — associated with a dry hole were expensable (i.e., deductible in the year in which the well was determined to be dry). Under the 2005 act, both integrated and independent producers were required to amortize the G&G component of the dry hole costs over two years. This reduced the incentive for G&Gs associated with a dry hole but increased the incentive for G&Gs associated with most successful wells. This provision became effective for G&G amounts paid or incurred in taxable years beginning after the date of enactment.

Two-year amortization of G&G costs is still allowed for independent producers, but as a result of a provision in the Tax Increase Prevention and Reconciliation Act (P.L. 109-222, enacted in May 2006), integrated producers must now amortize such costs over five years.[15] Amortization means that the costs are deducted evenly — the same absolute dollars are taken as deductions every year over a specified period of time, in this case two or five years. It is also called straight-line depreciation.[16]

Determination of Independent Producer Status for Purposes of the Oil Depletion Deduction

Firms that extract oil, gas, or other minerals are permitted a deduction to recover their capital investment in a mineral reserve, which depreciates due to physical and economic depletion or exhaustion as the mineral is recovered (IRC § 611). Depletion, like depreciation, is a form of capital recovery: an asset, the mineral reserve itself, is being expended to produce income. Under the income tax, such a loss in value or cost is deductible.

There are two methods of calculating this deduction: cost depletion and percentage depletion. Cost depletion allows for the recovery of the actual capital investment — the costs of discovering, purchasing, and developing a mineral reserve. Each year, and over the period during which the reserve produces income, the taxpayer deducts a portion of the adjusted basis (original capital investment less previous deductions) equal to the fraction of the estimated remaining recoverable reserves that have been extracted and sold. Under this method, the total deductions cannot exceed the original capital investment.

Under percentage depletion, the deduction for recovery of capital investment is a fixed percentage as set by law of the "gross income" (i.e., revenue) from the sale of the mineral. Under this method, total deductions typically exceed, despite the limitations, the capital invested to acquire and develop the reserve.

IRC § 613 states that mineral producers must claim the higher of cost or percentage depletion. The percentage depletion rate for oil and gas is 15% and is limited to average daily production of 1,000 barrels of oil, or its equivalent in gas. For producers of both oil and gas, the limit applies on a combined basis. For example, an oil-producing company with 2006 oil production of 100,000 barrels and natural gas production of 1.2 billion cubic feet (the statutory equivalent of 200,000 barrels of oil) has average daily production of 821.92 barrels (300,000 ÷ 365 days).

Percentage depletion is not available to integrated major oil companies; it is available only for independent producers and royalty owners.

Beginning in 1990, the percentage depletion rate was raised on production from marginal wells — oil from stripper wells (those producing no more than 15 barrels per day, on average) and heavy oil. This rate starts at 15% and increases by one percentage point for each whole $1 that the reference price of oil for the previous calendar year is less than $20 per barrel (subject to a maximum rate of 25%). This higher rate is also limited to independent producers and royalty owners, and for up to 1,000 barrels, determined as before on a combined basis (including non-marginal production). Small independents operate nearly 400,000 small stripper wells in about 28 states, about 78% of the nearly 510,000 producing wells in the United States. Output from stripper wells represented about 16% of total domestic production (about 850,000 barrels per day) in the United States in 2004.[17]

The percentage depletion deduction is limited to 65% of the taxable income from all properties for each producer. A second limitation, the 100% net-income limitation, which applied to each individual property rather than to all the properties, was retroactively suspended for oil and gas production from marginal wells by the Working Families Tax Relief Act of 2004 (P.L. 108-311) through December 31, 2005. The 100% net-income limitation also had been suspended from 1998 to 2003. The difference between percentage depletion and cost depletion is considered a subsidy. It was once a tax preference item for purposes of the alternative minimum tax, but this was repealed by the Energy Policy Act of 1992 (P.L. 102-486).

The percentage depletion allowance is available for other types of fuel minerals, at rates ranging from 10% (coal, lignite) to 22% (uranium), and for mined hard rock minerals. The rate for regulated natural gas and gas sold under a fixed contract is 22%; the rate for geo-pressurized methane gas is 10%. Oil shale and geothermal deposits qualify for a 15% allowance. The net-income limitation to percentage depletion for coal and other fuels is 50%, compared with 100% for oil and gas. Under code section 291, percentage depletion on coal mined by corporations is reduced by 20% of the excess of percentage over cost depletion.

For purposes of percentage depletion, before EPACT05, an independent oil producer was one that, on any given day, (1) did not refine more than 50,000 barrels of oil and (2) did not have a retail operation grossing more than $5 million a year (IRC § 613A[d]). EPACT05 raised the 50,000 barrel daily limit to 75,000. In addition, the act changed the refinery limitation from actual daily production to average daily production for the taxable year. Accordingly, the average daily refinery runs for the taxable year may not exceed 75,000 barrels. For this purpose, the taxpayer would calculate average daily refinery runs by dividing total refinery runs for the taxable year by the total number of days in the taxable year. This is effective for taxable years ending after the date of enactment.

Natural Gas Distribution Lines Treated as 15-Year Property

For purposes of determining the depreciation deduction, EPACT05 established a 15-year recovery period for natural gas distribution lines. Prior to this amendment, natural gas distribution lines were assigned a 20-year recovery period. This provisions is effective for property, the original use of which begins with the taxpayer after April 11, 2005, which is placed in service after April 11, 2005, and before January 1, 2011, and does not apply to property subject to a binding contract on or before April 11, 2005.

Temporary Expensing for Equipment Used in Oil Refining

Before the enactment of EPACT05, depreciation rules (the Modified Accelerated Cost Recovery System, MACRS) required oil refinery assets to be depreciated over 10 years using the double declining balance method.[18] Under the 2005 act, refineries are allowed to irrevocably elect to expense 50% of the cost of qualified refinery property, with no limitation on the amount of the deduction. This provision was enacted to increase investments in existing refineries so as to increase petroleum product output and reduce prices.

The expensing deduction is allowed in the taxable year in which the refinery is placed in service. The remaining 50% of the cost remains eligible for regular cost recovery provisions. To qualify for the deduction (1) original use of the property must commence with the taxpayer; (2)(a) construction must be pursuant to a binding construction contract entered into after June 14, 2005, and before January 1, 2008, (b) in the case of self-constructed property, construction began after June 14, 2005, and before January 1, 2008, or (c) the refinery is placed in service before January 1, 2008; (3) the property must be placed in service before January 1, 2012; (4) the property must meet certain production capacity requirements if it is an addition to an existing refinery; and (5) the property must meet all applicable environmental laws when placed in service. Certain types of refineries, including asphalt plants, are not eligible for the deduction, and there is a special rule for sale-leasebacks of qualifying refineries. If the owner of the refinery is a cooperative, it may elect to allocate all or a part of the deduction to the cooperative owners, allocated on the basis of ownership interests. This provision is effective for qualifying refineries placed in service after date of enactment (i.e., it became effective on August 9, 2005).

Arbitrage Rules Not To Apply to Prepayments for Natural Gas

EPACT05 creates a safe harbor exception to the general rule that tax-exempt, bond-financed prepayments violate the tax code's arbitrage restrictions. The term *investment-type property* does not include a prepayment under a qualified natural gas supply contract. The act also provides that such prepayments are not treated as private loans for purposes of the private business tests. Thus, a prepayment financed with tax-exempt bond proceeds for the purpose of obtaining a supply of natural gas for service area customers of a governmental utility would not be treated as the acquisition of investment-type property. The safe harbor provisions do not apply if the utility engages in intentional acts to render (1) the volume of natural gas covered by the prepayment to be in excess of that needed for retail natural gas consumption and (2) the amount of natural gas that is needed to fuel transportation of the natural gas to the governmental utility. This provision is effective for obligations issued after date of enactment.

Natural Gas Gathering Lines Treated as Seven-Year Property

Under tax law prior to the enactment of EPACT05, the recovery period for natural gas gathering lines could be either 7 or 15 years, depending on whether they were classified as production or transportation equipment. Several court cases reflected the ambiguous tax treatment. Natural gas pipelines had a recovery period of 15 years, whereas natural gas distribution lines had a recovery period of 20 years (which, as noted above, was reduced to 15 years). EPACT05 assigned natural gas gathering lines a seven-year recovery period for MACRS depreciation deductions.

EPACT05 defined a natural gas gathering line as the pipe, equipment, and appurtenances determined to be a gathering line by the Federal Energy Regulatory Commission (FERC) or used to deliver natural gas from the well-head or common point to the point at which the gas first reaches (1) a gas processing plant, (2) an interconnection with an interstate transmission line, (3) an interconnection with an intrastate transmission pipeline, or (4) a direct connection with a local distribution company, a gas storage facility, or an industrial consumer. Also, the act requires that the original use of the property begin with the taxpayer. This provision became effective for property placed in service after April 11, 2005, excluding property with respect to which the taxpayer or related party had a binding acquisition contract on or before April 11, 2005.

Pass Through to Owners of Deduction for Capital Costs Incurred by Small Refiner Cooperatives in Complying with EPA Sulfur Regulations

IRC § 45H allows a small refiner to claim a tax credit for the production of low- sulfur diesel fuel that is in compliance with Environmental Protection Agency (EPA) sulfur regulations (the Highway Diesel Fuel Sulfur Control Requirements). The credit is $2.10 per barrel of low-sulfur diesel fuel produced; it is limited to 25% of the capital costs incurred by the refiner to produce the low-sulfur diesel fuel. The 25% limit is phased out proportionately as a refiner's capacity increases from 155,000 to 205,000 barrels per day.

Section 179B allows a small refiner to also claim a current year tax deduction (i.e., expensing), in lieu of depreciation, for up to 75% of the capital costs incurred in producing low-sulfur diesel fuel that is in compliance with EPA sulfur regulations. This incentive is also prorated for refining capacity between 155,000 and 205,000 barrels per day. The taxpayer's basis in the property that receives the exemption is reduced by the amount of the production tax credit. In the case of a refinery organized as a cooperative, both the credit and the expensing deduction may be passed through to patrons.

For both incentives, a small business refiner is a taxpayer who (1) is in the business of refining petroleum products, (2) employs not more than 1,500 employees directly in refining, and (3) has less than 205,000 barrels per day (averaged over the year) of total refining capacity. The incentives took effect retroactively beginning on January 1, 2003.

EPACT05 provided that cooperative refineries that qualify for § 179B expensing of capital costs incurred in complying with EPA sulfur regulations could elect to allocate all or part of the deduction to their owners, determined on the basis of their ownership interests. The election is made on an annual basis and is irrevocable once made. The provision became effective as if included in § 338(a) of the American Jobs Creation Act of 2004, which introduced the tax credit.

Modification and Extension of Credit for Producing Fuel from a Nonconventional Source for Facilities Producing Coke or Coke Gas[19]

Section 45K of the Internal Revenue Code (IRC) provides for a production tax credit of $3 per barrel of oil-equivalent (in 1979 dollars) for certain types of liquid, gaseous, and solid fuels produced from selected types of alternative energy sources (so-called "non-conventional fuels") and sold to unrelated parties. The full credit is available if oil prices fall below $23.50 per barrel (in 1979 dollars); the credit is phased out as oil prices rise above $23.50 (in 1979 dollars) over a $6 range (i.e., the inflation-adjusted $23.50 plus $6).

Both the credit and the phase-out ranges are adjusted for inflation (multiplied by an inflation adjustment factor) since 1979. With an inflation adjustment factor of 2.264 (meaning that prices, as measured by the Gross Domestic Product deflator, have more than doubled since 1979), the credit for 2005 production was

$6.79 per barrel of oil equivalent, which is the amount of the qualifying fuel that has a British Thermal Unit (Btu) content of 5.8 million. The credit for gaseous fuels was $1.23 per thousand cubic feet (mcf). The credit for tight sands gas is not indexed to inflation; it is fixed at the 1979 level of $3 per barrel of oil equivalent (about $0.50 per mcf). In 2005, the reference price of oil, which was $50.76 per barrel, still below the inflation adjustment phase-out threshold oil price of $53.20 for 2005 ($23.50 multiplied by 2.264), the full credit of $6.56 per barrel of equivalent was available for qualifying fuels.

Qualifying fuels include synthetic fuels (liquid, gaseous, and solid) produced from coal, and gas produced from either geopressurized brine, Devonian shale, tight formations, or biomass. To qualify for the credit, synthetic fuels from coal must undergo a significant chemical transformation, defined as a measurable and reproducible change in the chemical bonding of the initial components. In most cases, producers apply a liquid bonding agent to the coal or coal waste (coal fines), such as diesel fuel emulsions, pine tar, or latex, to produce a solid synthetic fuel. The coke made from coal and used as a feedstock, or raw material, in steel-making operations also qualifies as a synthetic fuel, as does the breeze (small pieces of coke) and the coke gas (produced during the coking process). Depending on the precise Btu content of these synfuels, the § 45K tax credit could be as high as $26 per ton or more, which is a significant fraction of the market price of coal. Qualifying fuels must be produced within the United States. The credit for coke and coke gas is also $3 per barrel of oil equivalent and is also adjusted for inflation, but the credit is set to a base year of 2004, making the nominal unadjusted tax credit less than for other fuels.

The section 45K credit for gas produced from biomass, and synthetic fuels produced from coal or lignite, is available through December 31, 2007, provided that the production facility was placed in service before July 1, 1998, pursuant to a binding contract entered into before January 1, 1997. The credit for coke and coke gas is available through December 31, 2009, for plants placed in service before January 1, 1992, and after June 30, 1998. The section 45K credit used to apply to oil produced from shale or tar sands, and coalbed methane (a colorless and odorless natural gas that permeates coal seams and that is virtually identical to conventional natural gas). However, the credit for these fuels terminated on December 31, 2002 (and the facilities had to have been placed in service, or wells drilled, by December 31, 1992).

The section 45K credit is part of the general business credit. It is not claimed separately; it is added together with several other business credits and is also subject to the limitations of that credit. The section 45K credit is offset (or reduced) by certain other types of government subsidies that a taxpayer may benefit from: government grants, subsidized or tax-exempt financing, energy investment credits, and the enhanced oil recovery tax credit that may be claimed with respect to such projects. Finally, the credit is nonrefundable and cannot be used to offset a taxpayer's alternative minimum tax liability. Any unused section 45K credits generally may not be carried forward or back to another taxable year. (However, under the minimum tax section 53, a taxpayer receives a credit for prior-year minimum tax liability to the extent that a section 45K credit is disallowed as a result of the operation of the alternative minimum tax.)

The Energy Policy Act of 2005 made several amendments to the section 45K tax credit. First, the credit's provisions were moved from § 29 of the tax code to new § 45K. Before this, this credit was commonly known as the "section 29 credit." Second, the credit was made available for qualified facilities that produce coke or coke gas that were placed in service before January 1, 1993, or after June 30, 1998, and before January 1, 2010. Coke and coke gas produced and sold during the period beginning on the later of January 1, 2006, or the date the facility is placed in service, and ending on the date which is four years after such period begins, are eligible for the production credit, but at a reduced rate and only for a limited quantity of fuel. The tax credit for coke and coke gas is $3.00 per barrel of oil equivalent, but the credit is indexed for inflation starting with a 2004 base year, compared with a 1979 base year for other fuels. A facility producing coke or coke gas and receiving a tax credit under the previous § 29 rules is not eligible to claim the credit under the new section 45K. The new provision also requires that the amount of credit-eligible coke produced not exceed an average barrel-of-oil equivalent of 4,000 barrels per day. Third, the 2005 act provided that with respect to the IRS moratorium on taxpayer-specific guidance concerning the credit, the IRS should consider issuing rulings and guidance on an expedited basis to those taxpayers who had pending ruling requests at the time that the IRS implemented the moratorium. Finally, the 2005 legislation made the general business limitations applicable to the tax credit. Any unused credits can be carried back one year and forward 20 years, except that the credit cannot be carried back to a taxable year ending before January 1, 2006. These new rules were made effective for fuel produced and sold after December 31, 2005, in taxable years ending after that date.

Revenue Effects

Table 1 shows the revenue effects of the tax provisions in EPACT05, organized by type of incentive. These are the original revenue effects estimated for EPACT05, signed into law on August 8, 2005, by the Joint Committee on Taxation (JCT). Because of changes to energy prices, energy markets, and general economic conditions, revenue loss estimates of the same provisions calculated today would most likely differ from those original estimates.

JCT's estimated revenue losses were projected over an 11-year time frame, from FY2005 to FY20 15. The total revenue losses are reported in two ways: the absolute dollar value of tax cuts over 11 years, and the percentage distribution of total revenue losses by type of incentive. Each of the seven tax subsidies for the oil and gas industry are shown separately, as well as the aggregate for upstream (exploration, development, and production) operations and downstream operations (refining and transportation/distribution). Also, for perspective, the oil and gas tax revenue losses are compared with those for other industries and with the tax subsidies for energy efficiency and alternative/renewable fuels.

The JCT estimates that the 2005 act provides about $2.6 billion in tax cuts for the oil and gas industry as a whole over 11 years, comprising about $1.1 billion for upstream operations and $1.5 billion for downstream, or refining and distribution, operations. For energy conservation and efficiency, the 2005 act provides about $1.3 billion, including a deduction for energy-efficient commercial property, fuel cells, and micro-turbines. Renewables incentives include a two-year extension of the tax code § 45 credit, renewable energy bonds, and business credits for solar. The total renewable tax subsidies in EPACT05 were about $4.5 billion.

Although the above oil and gas tax subsidies may not be justified based on economic theory, and considering the high oil and gas prices over much of the policy period, they are not large when measured relative to the industries' gross product, which measures in the hundreds of billions of dollars.[20] Another misconception is that industry was the beneficiary of many and significant tax breaks before these provisions were enacted. The industry did benefit historically from significant tax subsidies; however, most of these had been either eliminated or pared back since the 1970s.

Table 1. Energy Tax Provisions in the Energy Tax Act of 2005 (P.L. 109-58): 11-Year Estimated Revenue Loss, by Type of Incentive

	Amount ($ millions)	Percentage
Incentives for fossil fuels supply		
(1) Oil & Gas Production:	-1,132	7.8%
a) amortize all G&G costs over 2 years	-974	
b) liberalize the definition of independent producer	-158	
(2) Oil & Gas Refining and Distribution:	-1,501	10.4%
a) gas pipelines treated as 15-year property	-1,019	
b) temporary expensing in refining of liquid fuels	-406	
c) exempt prepayment of natural gas from arbitrage	-53	
d) gas gathering lines treated as 7-year property	-16	
c) expensing for coop refinery of low sulfur diesel	7	
(3) Coal	-2,948	20.4%
(4) Subtotal	-5,581	38.6%
Electricity restructuring provisions		
(5) Nuclear	-1,571	10.9%
(6) Other	-1,549	10.7%
(7) Subtotal	-3,120	21.6%
Incentives for efficiency, renewables, and alternative fuels		
(8) Energy Efficiency	-1,260	8.7%
(9) Renewable Energy & Alternative Fuels	-4,500	31.1%
(10) Subtotal	-5,760	39.8%
(11) Net Energy Tax Cuts	-14,461	100.0%
(12) Non Energy Tax Cuts[a]	-92	
(13) Total Energy and Non-Energy Tax Cuts	-14,553	
(14) Energy Tax Increases[b]	+2,857	
(15) Other Tax Increases	171	
(15) *Net tax cuts*	-11,525	

Source: CRS compilation based on Joint Committee of Taxation estimates.

[a] The act includes a provision to expand R&D for all energy activities. This provision is listed as a non energy tax cut to simplify the table.

[b] Energy tax increases comprise the oil spill liability tax and the Leaking Underground Storage Tank financing rate, both of which are imposed on oil refineries. If these taxes are subtracted from the tax subsidies (row 2), the oil and gas refinery and distribution sector received a net tax increase of $1,356 ($2,857-$1,501).

Table 2. Special Tax Incentives Targeted for the Oil and Gas Industry and Estimated Revenue Losses, FY2006

Category	Provision	Major Limitations	Original Enacting Legislation/ Regulation	Federal Revenue Losses FY2006($ millions)
Expensing of Intangible Drilling Costs (IDCs) and Amortization of Exploration and Development Expenses	Firms engaged in the exploration and development of oil or gas properties may expense (deduct in the year paid or incurred) rather than capitalize certain types of drilling expenditures. Geological and geophysical expenses paid or incurred in connection with the domestic exploration for, or development of, oil or gas can be amortized ratably (evenly) over five years.	Integrated oil/gas corporations may expense only 70% of IDCs; the remaining 30% must be amortized and all of the excess IDCs over the 10-year amortizable amount are subject to the alternative minimum tax.	1916 Treasury Regulation T.D. 45, article 223	1 ,100[a]
Percentage Depletion Allowance	Firms that extract oil or gas are permitted to deduct 15% of sales (up to 25% for marginal wells depending on oil prices) to recover their capital investment in a mineral reserve.	Percentage depletion is available only for independent producers (and royalty owners) and only up to 1,000 barrels or equivalent per day; it is limited to 100% of the net income from any individual property and to 65% of the taxable income from all properties for each producer.	Revenue Act of 1926	1,000
Incentives for Small	IRC § 45H allows a small refiner to claim a $2.10 credit	Credit limited to 25% of capital costs; expensing phases out for	P.L. 108-357	50[b]

Refiners to Comply with EPA Sulfur Regulations	per barrel of low-sulfur diesel produced that complies with EPA sulfur regulations. IRC§ 179B allows a small refiner to expense, in lieu of depreciation, up to 75% of the capital costs incurred in producing low-sulfur diesel fuel that is in compliance with EPA sulfur regulations.	refining capacity of 155,000-205,000 barrels per day.		
Category	Provision	Major Limitations	Original Enacting Legislation/ Regulation	Federal Revenue Losses FY2006($ millions)
Tax Credits for Enhanced Oil Recovery Costs	IRC § 43 provides for a 15% income tax credit for the costs of recovering domestic oil by qualified "enhanced oil recovery" (EOR) methods, to extract oil that is too viscous to be extracted by conventional primary and secondary water-flooding techniques.	The EOR credit is nonrefundable and is allowable provided that the average wellhead price of crude oil (using West Texas Intermediate as the reference), in the year before credit is claimed, is below the statutorily established threshold price of $28 (as adjusted for inflation since 1990), in the year the credit is claimed. With average wellhead oil prices for 2005 (about $65) well above the reference price (about $38) the EOR credit was not available.	P.L. 10 1-508	0

Table 2. (Continued)

Marginal Production Tax Credit	A $3 tax credit is provided per barrel of oil ($0.50/thousand cubic feet [mcf]) of gas from marginal wells, and for heavy oil.	The credit phases out as oil prices rise from $15 to $18 per barrel (and as gas prices rise from $1.67 to $2.00/thousand cubic feet), adjusted for inflation. The credit is limited to 25 barrels per day or equivalent amount of gas and to 1,095 barrels per year or equivalent. Credit may be carried back up to five years. At 2005 oil and gas prices, the marginal production tax credit was not available.	P.L. 108-357	0

Source: Joint Tax Committee estimates and Internal Revenue Service data.

[a] The revenue loss estimate excludes the benefit of expensing costs of dry tracts and dry holes, which includes expensing some things that would otherwise be capitalized. This is a normal feature of the tax code but confers special benefits on an industry where the cost of finding producing wells includes spending money on a lot that turn out dry. The revenue loss estimates also include revenue losses associated with the passive loss limitation rule exemption for the oil and gas industry.

[b] The JCT reports this revenue loss at less than $50 million but does not report the actual figure.

Tax Increases

Subtitle F of EPACT05 describes the four tax increases or revenue offsets. Two of the tax increases — modification of the § 197 amortization, and an increase in the excise taxes on tires — are negligible, raising taxes by just under an estimated $200 million over 11 years. However, the other two are sizeable tax increases for the oil and gas industry: reinstatement of the Oil Spill Liability Trust Fund and extension of the Leaking Underground Storage Tank (LUST) trust fund rate, which would be expanded to all fuels.

The total oil and gas industry tax increases are roughly $2.8 billion over 11 years, for a net increase in taxes on the industry of about $200 million, according to the JCT estimates. However, because the oil spill liability tax and the Leaking Underground Storage Tank financing taxes are excise taxes on oil and petroleum products, and are imposed on oil refineries, the net effect of the 2005 act on the oil and gas refinery sector was a tax increase of about $1.3 billion over 11 years.

OTHER OIL AND GAS TAX SUBSIDIES

The Energy Policy Act of 2005 expanded some (but not all) of the preexisting tax subsidies for oil and gas and introduced several new ones. Thus, some of the recent proposals to roll back tax subsidies to oil and gas focus on the subsidies that were in effect before the 2005 act, and which continue be in effect.

OTHER OIL AND GAS TAX SUBSIDIES

A list of the preexisting federal tax subsidies (incentives) available for the U.S. oil and gas industry — those in effect before EPACT05 and still in effect today — (and their corresponding revenue loss estimates) appears in Table 2. The corresponding revenue losses, as estimated by the JCT in its latest tax expenditures compendium, appear in the last column.[21] Note that the table defines tax subsidies or incentives targeted for the oil and gas industry as those that are due to provisions in the tax law that apply only to this industry and not to others.

General Tax Provisions that May Benefit the Oil and Gas Industry

This discussion has so far excluded current-law tax provisions and incentives that may apply to non-oil and gas businesses but that may also confer tax benefits to the oil and gas industry. There are numerous such provisions in the tax code, which some have called loopholes — they are not strictly considered to be tax expenditures. A complete listing of them is beyond the scope of this report; however, four examples, which have been under discussion as possible revenue raisers, follow to illustrate the point.

For example, the current system of depreciation generally allows the writeoff of equipment and structures somewhat faster than would be the case under both general accounting principles and economic theory; the JCT treats the excess of depreciation deductions over the alternative depreciation system as a tax subsidy (or tax expenditure). In FY2006, the JCT estimates that the aggregate economy-wide revenue loss from this accelerated depreciation deduction (including the expensing under IRC § 179) is $6.7 billion. A certain, but unknown, fraction of this revenue loss or tax benefit accrues to the domestic oil and gas industry, but separate estimates are unavailable.

A second example is the deduction for domestic production (or manufacturing) activities under IRC § 199, which, as noted above is the target of H.R. 5218 (109th Congress). Enacted under the American Jobs Creation Act of 2004 (P.L. 108-357, also known as the JOBS bill), the domestic production deduction (IRC § 199) generally allows taxpayers to receive a deduction based on qualified production activities income resulting from domestic production. The deduction is 3% of income for 2006, rising to 6% between 2007 and 2009, and 9% thereafter; it is subject to a limit of 50% of the wages paid that are allocable to domestic production during the taxable year. The revenue impact of this provision is anticipated by the JCT to be a loss of $4.8 billion of federal revenue in FY2007, and $76 billion over the first 10 years of its life. A certain (as yet unknown) fraction of the tax benefits from the deduction will accrue to the domestic oil and gas industry. The deduction applies to oil and gas or any primary product thereof, provided that such product was "manufactured, produced, or extracted in whole or in significant part in the United States." Recently, the JCT estimated the revenues that would be gained by repealing this deduction for the domestic oil and gas industry at about $0.2 billion in FY2007, and about $2 billion from FY2007-FY20 12.[22]

A third example concerns the "last-in/first-out" (LIFO) system of inventory accounting under IRC § 472. This method values the goods sold as the most recent inventory purchase. During a period of rising prices, this method of inventory accounting increases production costs and reduces taxable income and tax liabilities. A provision in the Senate version of H.R. 4297 (109th Congress) would have eliminated a portion of the tax benefits from LIFO inventory accounting for major integrated oil companies with gross receipts in excess of $1 billion. Under threat of presidential veto, this provision, which would have increased taxes on such companies by an estimated $3.5 billion in FY2006, was deleted from the final law, the Tax Increase Prevention and Reconciliation Act of 2006 (P.L. 109-222).[23]

A fourth example is the foreign tax credit, which is a federal tax credit against U.S. tax liabilities for *income* taxes paid to foreign countries. This section of the tax code is intended to prevent the double taxation of foreign source income (income earned abroad by U.S. residents and corporations). However, many countries in which domestic U.S. oil companies conduct business (either through branches or foreign subsidiaries) impose levies that are not strictly considered to be creditable income taxes, which may have the effect of going beyond prevention of double taxation of foreign source income — it may actually lead to a reduction of taxes on domestic source income. A provision in the Senate version of H.R. 4297 (109th Congress) would have denied the foreign tax credit, under certain conditions, for major integrated oil companies with gross receipts in excess of $1 billion. The foreign tax credit would have been denied in the event that the foreign levy was assessed in exchange for an economic benefit provided by the foreign jurisdiction to the domestic oil company and if the foreign jurisdiction did not generally impose an income tax. This provision, which would have increased taxes on such companies by an estimated $0.8 billion over the 10-year period from FY2006 to FY2015, was deleted from the final law, the Tax Increase Prevention and Reconciliation Act of 2006 (P.L. 109-222).[24]

Finally, Table 2 excludes targeted taxes that impose special tax liabilities on the domestic oil and gas industry — taxes that are not imposed on other industries. These would include taxes such as the motor fuels excise taxes (e.g., the 1 8.4¢ per gallon tax on gasoline, the 24.4¢ per gallon tax on diesel) and the oil spill liability trust fund excise tax, which imposes a $0.05 per barrel tax on every barrel of crude oil refined domestically.[25] These taxes are imposed on refiners, although under normal (and stable) market conditions they are shifted forward (or passed through the distribution and retailing chain) and largely paid by consumers. The motor fuels excise taxes (including the Leaking Underground Storage Tank Trust Fund Tax) represent a tax liability — the amount of revenues collected by the federal government — of about $36 billion in FY2006;[26] revenues collected from the oil spill liability excise tax are estimated by the JCT at $0. 150 billion.

ENDNOTES

[1] U.S. Congress. Congressional Budget Office. H.R. 6, CLEAN Energy Act of 2007. (Letter to Chairman Nick Rahall, Committee on Natural Resources.) Jan. 12, 2007. 4 p. [http://www.cbo.gov/ftpdocs/77xx/doc7728/hr6prelim.pdf]

[2] For a summary and analysis of this law, see CRS Report RL33302, *Energy Policy Act of 2005: Summary and Analysis of Enacted Provisions*. The two-year amortization period was slowed down to five years for integrated producers under 2006 tax legislation, as discussed in the text.

[3] As discussed later, these subsidies were largely eliminated on much of the oil production and assets, but other, less significant subsidies — the special exemption from the passive loss limitation rules and some special tax credits — were added to the tax code.

[4] Glasner, David, *Politics, Prices, and Petroleum: The Political Economy of Energy*, Pacific Institute for Public Policy Research, 1985, pp. 142-144.

[5] A stripper well is one that produces small quantities of oil and natural gas. The tax law currently defines this limit as 15 barrels of oil or the equivalent amount of natural gas per day; the oil and gas industry defines it 10 barrels per day or less.

[6] Although no tax bill was passed that reduced taxes on oil and gas, the 106th Congress did enact a package of $500 million in loan guarantees for small independent producers, which became law (P.L. 106-5 1), in August 1999.

[7] After some existing energy tax incentives expired in 2003, the 108th Congress enacted retroactive extension of several of the provisions as part of the Working Families Tax Relief Act of 2004 (P.L. 108-311). That law, which reduced revenues by about $1.3 billion over 10 years, was enacted on October 4, 2004. About $5 billion in energy tax incentives — both expansion or liberalization of some of the more popular energy tax provisions, as well as some new energy tax incentives — were part of the American Jobs Creation Act of 2004 (P.L. 108-357) enacted on October 22, 2004.

[8] For an analysis of the windfall profit tax, see CRS Report RL33305, *The Crude Oil Windfall Profits Tax of the 1980s: Implications for Current Energy Policy*, by Salvatore Lazzari.

[9] Prior to the 2005 act, G&G costs for dry holes were expensed in the first year and capitalized for successful wells.

[10] Bureau of National Affairs, Daily Tax Report. *House Democratic Leadership Letter to Speaker Hastert Asking for Rollback of Tax Breaks for Oil Companies*, Sept. 5, 2006.

[11] As discussed below, many of the remaining tax subsidies are available only to independent oil and gas producers, which, however, may be very large.

[12] There is an important economic distinction between a subsidy and a tax benefit. As is discussed elsewhere in this report, firms receive a variety of tax benefits that are not necessarily targeted subsidies (or tax expenditures) because they are available generally.

[13] McKinnon, John D. "Are Higher Taxes in the Offing?" *The Wall Street Journal*, Oct. 30, 2006, p. A-6; Bureau of National Affairs, "Menu of Proposals Available to Democrats Looking to Roll Back Oil, Energy Tax Breaks," *Daily Tax Report*, Nov. 14, 2006, p. G-2.

[14] These are CRS compilations based on Joint Committee on Taxation estimates. See U.S. Congress, Joint Committee on Taxation, *Estimated Budget Effects of the Conference Agreement for Title XIII of H.R. 6, The "Energy Tax Incentives Act of 2005,"* July 27, 2005, JCX-59-05.

[15] The 2006 amendment constitutes a reduction in the tax benefits and was part of the compromise for allowing the G&G costs of successful wells to be amortized over two years rather than capitalized.

[16] The term *amortization* is also used in tax parlance as referring to the depreciation of intangible property, such as patents and copyrights.

[17] Both the number of stripper wells and oil output from such wells is reported in American Petroleum Institute, *Basic Petroleum Data Book*, vol. 26, no. 2, (section IV, table 3), August 2006.

[18] Under the double declining balance method of calculating depreciation deductions, the annual deduction is a fixed percentage (200% or double the straight-line rate) of the difference between asset cost and prior year depreciation deductions.

[19] Two of the nine special tax subsidies for oil and gas in EPACT05 were for unconventional gases and synfuels from coal under the § 45K tax credit. These provisions are discussed because the § 45K tax credit has been important to the development of unconventional gases such as coalbed methane and tight sands gas. However, its revenue losses are subsumed under the coal category of Table 1 largely because in recent years the provision has benefitted primarily the coal industry by increasing the demand for coal.

[20] For the economic theory of taxation of exhaustible natural resources, see CRS Report RL30406, *Energy Tax Policy: An Economic Analysis*, by Salvatore Lazzari.

[21] U.S. Congress, Joint Committee Print, *Estimates of Federal Tax Expenditures for Fiscal Years 2006-2010*, prepared for the House Committee on Ways and Means and the Senate Committee on Finance by the Joint Committee on Taxation Staff, Apr. 25, 2006.

[22] U.S. Congress, Joint Committee on Taxation, *JCT Cost Estimate for McDermott-Kerry Legislation (H.R. 5218, S. 2672) to Eliminate Oil Company Eligibility for JOBS Act Section 199 Tax Breaks,* May 10, 2006.

[23] U.S. Congress. Joint Committee on Taxation. *Comparison of Estimated Revenue Effects of the Tax Provisions Contained in H.R. 4297, "The Tax Relief Extension Reconciliation Act of 2005," As Passed by the House, and H.R. 4297, "The Tax Relief Act of 2005, " As Passed by the Senate*. February 9, 2006.

[24] U.S. Congress. Joint Committee on Taxation. *Comparison of Estimated Revenue Effects of the Tax Provisions Contained in H.R. 4297, "The Tax Relief Extension Reconciliation Act of 2005," As Passed by the House, and H.R. 4297, "The Tax Relief Act of 2005, " As Passed by the Senate*. February 9, 2006.

[25] Moneys are allocated into a fund for cleaning up oil spills.

[26] Revenues from motor fuels excise taxes are allocated primarily to the Highway Trust Fund (HTF) and various trust funds, depending on the mode of transportation. The HTF also includes revenue from excise taxes on tires, a heavy vehicle use tax, and retail sales tax on trucks and tractors.

In: The Oil Industry and Windfall Profits
Editor: Maria B. Coswell

ISBN 978-1-60692-840-0

Chapter 4

THE USE OF PROFIT BY THE FIVE MAJOR OIL COMPANIES*

Robert Pirog

SUMMARY

The price of crude oil began to increase in the last quarter of 2003, and has led to the high prices observed from 2004 through 2007. The Iraq War, unexpectedly high demand growth in China, India, and the United States, and Hurricanes Katrina and Rita, along with a number of other factors, all contributed to the rising price.

An important result of these largely unexpected events was that the oil industry, as represented by the five major integrated oil companies doing business in the United States, experienced rapidly expanding revenues and profits. Some observers characterized these profits as "windfall" gains, while others pointed to the increasing scarcity and rising costs observable in the oil industry. Some saw "price gouging" in high gasoline prices, while others saw the market working to avoid physical supply shortages.

The larger profits experienced by the oil industry and the five major integrated oil companies can be used in a variety of ways. Profits might be used to expand exploration and development of crude oil resources to expand the supply

* This is an edited, excerpted and augmented edition of a CRS Report for Congress publication, Order Code RL34044, dated April 11, 2008.

of oil. Refineries might be constructed, and technology improved at existing refineries, to expand the supply of petroleum products, most notably, gasoline. Profits might also be used to provide increased returns to the owners of the oil companies, the shareholders. This end might be accomplished through dividend payments and share repurchase plans. Profits might also be used to improve the balance sheets of the companies through debt reduction, potentially improving their financial health should they face a downturn in the market in the future. Until the effects of corporate plans that reflect a market characterized by higher oil prices can be observed, profits might tend to build up as cash reserves, as experienced by some of the five firms since 2004.

How the profits generated over the past four years are used will help to determine whether oil and petroleum product markets remain tight with high prices, or whether they loosen, develop extra capacity, and lead to moderating prices.

INTRODUCTION

The five major integrated oil companies operating in the U.S. market have earned net incomes totaling $424 billion since 2004.[1] In the previous four years, from 2000 to 2003, they earned net incomes of $171 billion.[2] This 148% increase in profit has attracted public attention and raised the issue of whether "windfall" profits had accrued to the firms. At the same time that these oil companies were earning increased profits, U.S. gasoline consumers were facing prices that rose above $3.30 per gallon, raising concerns that the increased profits might be tied to price gouging by the oil companies.

This report analyzes the uses of accrued profits by the five major integrated oil companies from 2004 through 2007.[3] Although the oil industry is composed of thousands of firms involved in many different aspects of the business, these five firms represent the visible face of the oil industry to the American public. These companies also earned 90% of the total earnings of integrated oil companies, and 74% of the earnings of all the integrated oil companies, the independent oil and gas producers, and the independent refiners and marketers in 2007. Because of their size, the decisions they make with respect to utilizing profits will largely determine how the industry's use of profit is viewed by the public.

How the industry uses its recent profits is important because the demand and supply balance, and hence petroleum product prices for the U.S. consumer, will be affected by the decisions made. For example, one of the commonly cited reasons for high gasoline prices in 2007 is that refinery capacity has been offline as a result of catching up on deferred maintenance and of accidents. If investments had been made in new refineries over the past 20 years, the balance between refinery capacity and product demand might not be as tight, thus reducing the pressure on gasoline prices.

Oil Prices and Profits

A primary source of the increased profits of the five oil companies has been the increased price of crude oil on the world market. The increased price of crude oil since 2004 has been attributed to the growth in demand from China, India, the United States, and other areas, as well as to hurricanes Katrina and Rita and a variety of other factors.[4] Few if any of these factors could be influenced by the five companies.

Crude oil prices began their rise toward the end of 2003, and although volatile, have remained at, or near, historically high levels since then.

Table 1. U.S. Composite Cost of Crude Oil to Refiners (dollars per barrel)

Year	2000	2001	2002	2003	2004	2005	2006	2007
Price	28.26	22.95	24.10	28.53	36.98	50.24	60.23	67.93

Source: Energy Information Administration, at [http://www.eia.doe.gov].

Note: The composite cost of crude oil represents the average price U.S. refiners pay for oil, and is closely related to petroleum product prices.

All five of the companies produce crude oil. Over the period 2004-2007, ExxonMobil produced an average of 2.6 million barrels per day (b/d), and was the leading producer among the five companies. Shell averaged 2.0 million b/d, BP averaged 2.5 million b/d, Chevron averaged 1.7 million b/d, and ConocoPhillips averaged 1.0 million b/d. Higher world prices for crude oil increased the revenue from this production in proportion to the increase in prices.

In addition to the increasing price of crude oil, tightness in the refining industry contributed to the increase in petroleum product prices, notably gasoline. All five of the selected companies are active in the refining industry. In 2007, these five companies accounted for approximately 38% of total U.S. refining capacity.

Although, historically, refining margins and profitability have tended to be volatile over time, profit margins have been at, or near, historic highs since 2004. During the fourth quarter of 2007 refining margins declined as the price of crude oil rose, and refiners were unable to quickly pass cost increases on to consumers because of the weakening demand growth of gasoline.

For the five companies, operating in both the upstream (exploration and production) and downstream (refining, distribution, and marketing) segments of the oil market has led to growing profitability, as shown in Table 3.

Table 2. Refining Capacity of Major Integrated Oil Companies (millions of barrels per day)

	2000	**2001**	**2002**	**2003**	**2004**	**2005**	**2006**	**2007**
ExxonMobil	1.77	1.79	1.80	1.84	1.84	1.86	1.86	1.86
BP	1.66	1.56	1.50	1.50	1.50	1.47	1.47	1.46
Shell	——	——	——	.59	.59	.53	.57	0.53
Chevron	1.04	1.05	1.08	1.00	1.00	1.01	1.01	1.01
Conoco Phillips	.938	2.25	2.27	2.19	2.19	2.29	2.22	1.78
Total	**6.44**	**6.65**	**6.65**	**7.13**	**7.13**	**7.16**	**7.13**	**6.64**

Source: American Petroleum Institute, *Basic Petroleum Data Book,* Section VIII, Table 11, February 2007, and Energy Information Administration.

Note: Shell data is for wholly owned refining capacity only. Shell also has a joint venture with Saudi Refining, Inc., called Motiva Enterprises, that operated an average of more than 760,000 barrels per day of refining capacity from 2000-2007.

Profits declined for the five companies from 2000 through 2002, then doubled in 2003. Profits rose by 39% in 2004, 34% in 2005, 8% in 2006, and declined by 1.3% in 2007. However, the two companies whose profits declined in 2007 each experienced singular losses, ConocoPhillips taking a writedown as a result of Venezuela nationalizing its Orinoco Basin investments, and BP experiencing pipeline problems in Alaska and refinery accidents in Texas. The magnitude of the profits earned, as well as the rapidity with which they accrued, has created a challenge for the oil companies: how to best utilize these resources and meet the varied demands of shareholders and the public.

Table 3. Net Income of Major Integrated Oil Companies (billions of dollars)

	2000	2001	2002	2003	2004	2005	2006	2007
ExxonMobil	17.7	15.3	11.4	21.5	25.3	36.1	39.5	40.6
BP	9.3	8.4	5.6	12.4	15.4	19.3	22.2	17.3
Shell	12.7	10.5	9.4	12.7	19.2	26.2	25.4	27.6
Chevron	5.1	3.2	1.1	7.2	13.3	14.1	17.1	18.7
Conoco Phillips	1.8	1.6	-0.29	4.7	8.1	13.5	13.5	11.9
Total	**46.6**	**39.0**	**27.2**	**58.5**	**81.3**	**109.2**	**117.7**	**116.1**

Source: Oil Daily, Profit Profile Supplements for covered years, company data, Securities and Exchange Commission filings.

Profit Allocation

Private corporations, such as the five major integrated oil companies, operate for the purpose of maximizing shareholder value, that is, to enhance, as much as possible, the value of the shares held by investors and the returns earned by those shares. The goal of maximizing shareholder value can be achieved in various ways. The management may choose to reinvest profits in the business, deploying new technology and capital equipment or hiring more workers. The management might also acquire assets for the corporation through merger or acquisition. Another strategy might be to directly pay cash dividends to the shareholders, or to buy back the company's shares on the open market to enhance the price of outstanding shares. The management also might decide to alter the capital structure of the company by reducing the outstanding debt of the company. Finally, the management might decide to hold the profits as cash, or other short-term assets, to acquire the flexibility to implement resource allocation decisions in the future. Of course, the company may apply the profits to several or all of these uses.

In principle, the expected rate of return on investments should be at least as great as the current rate of return earned by the company. If the management cannot identify investment opportunities that have expected rates of return in excess of the current rate of return, the profit is typically returned to the shareholders in the form of dividends or other payments.

Capital Expenditures

Many capital investment expenditures in the oil industry are allocated to projects that cost billions of dollars and will likely be online for decades. For example, an efficiently scaled refinery is likely to cost between $3 billion and $5 billion and operate for more than 30 years. When planning and investing in such facilities, the underlying variables that determine potential profits must not only be favorable now, but must also be forecast to be favorable for decades to come. Because of the magnitude of the funding required for such an investment, a mistake might cause damage to the company for years. These factors tend to create an investment philosophy in the oil industry that is characterized by a deliberate pace as well as a degree of conservatism in making capital expenditures.

These characteristics have been observable in the oil industry during the recent period of high profits. Some might have expected the increased price of oil to lead to an immediate boom in exploration and refinery construction. A slower pace of capital investment is consistent with a view that the currently high price of oil might decline in the future, leading to an investment reference price below the currently observable price, or a forecast that demand growth might slow or even decline.

Even if the decision had been quickly taken to invest in either upstream or downstream activities, there would likely be a lag between the decision to invest and substantial expenditures being made on the project. Many investment projects related to the oil industry require environmental permits from a number of federal, state, and local agencies, all of which might require studies to be undertaken and approvals to be granted. Several years in the permitting process might be expected.

Capital expenditures included in Table 4 might not be for only new or expanded capacity. Environmental regulations affect both the petroleum product slate as well as refinery sites, and may require capital investment to maintain compliance.

Table 4 shows that capital expenditures increased by less than 1/2% in 2004 compared to 2003, while net income rose by 39%. While capital expenditures rose by 17% from 2004 to 2005, profits rose by 34%. However, capital expenditures rose by 40% from 2005 to 2006, while profits rose by 8%. In 2007, capital expenditures rose by almost 8%, even though profits declined by 1.3%. The lag in capital expenditures might reflect a reassessment by the oil companies of future prices and profits, or it may reflect other factors that are less fundamental.

While it is possible that oil firms might want to invest both upstream and downstream to take advantage of favorable business conditions, their actions run the risk of reducing the potential profit of the market. Economic theory suggests that industries adjust to tight markets through relatively small increases in supply. Because the efficient scale of the oil industry tends to be large, investments tend to have a large enough effect on production levels, or capacity, to materially affect the market. For example, if only two of the five major firms decided to build new refineries, this might result in more than 1 million barrels a day of refining capacity coming online, an almost 6% increment, which could be enough to change the relative market balance.

Table 4. Capital Expenditures of Major Integrated Oil Companies (billions of dollars)

	2000	2001	2002	2003	2004	2005	2006	2007
ExxonMobil	8.4	9.9	11.4	12.8	11.9	13.8	15.5	15.4
BP	8.8	12.2	12.1	11.9	12.3	12.3	12.3	17.8
Shell	6.6	9.6	12.1	12.2	8.7	10.8	22.9	24.6
Chevron	3.6	9.7	7.6	5.6	6.3	8.7	13.8	16.7
Conoco Phillips	2.0	3.0	4.4	6.1	9.5	11.6	15.6	11.8
Total	**29.4**	**44.4**	**47.6**	**48.6**	**48.7**	**57.2**	**80.1**	**86.3**

Source: Securities and Exchange Commission Filing, forms 10-K and 20-F.

Mergers and Acquisitions

Depending on the relative prices of known oil resources and exploration, and existing facilities and construction, it has, on occasion, in the past, been cheaper for a company that wishes to expand to acquire assets financially, rather than through exploration and/or construction. This, along with other reasons, like scaling the company to an appropriate size for international competition, has led to periods of merger and acquisition activity in the oil industry.

Many of the major oil companies were involved in merger activity from 1998 to 2002. In 1999, Exxon and Mobil merged. BP followed its 1998 acquisition of Amoco with a takeover of Arco in 2000. Conoco and Phillips merged in 2002, and Chevron and Texaco combined in 2001.

Recent years have seen few transactions on the scale of 1998-2002. Chevron acquired Unocal in 2005 in a deal that was reported to cost $18 billion. ConocoPhillips acquired Burlington Resources in 2006 for a reported $35.5 billion.

Dividends and Stock Acquisition

If management has access to capital investment projects that are projected to earn more than the current rate of return for the company, carrying them out will increase the value of the company and can lead to capital appreciation for its outstanding shares. If such profitable projects cannot be identified, paying out dividends may be appropriate. Dividend payouts have positive and negative aspects. Shareholders may appreciate the immediate returns, but this payout is unlikely to lead to long-term capital appreciation of the shares. Taken in a positive light, the payout of dividends signals high earnings by the firm; taken in a negative light, they signal that management does not have wealth-increasing opportunities to use the funds to generate capital appreciation.

Table 5. Cash Dividends Paid by Major Integrated Oil Companies (billions of dollars)

	2000	**2001**	**2002**	**2003**	**2004**	**2005**	**2006**	**2007**
ExxonMobil	6.12	6.25	6.21	6.51	7.11	7.47	7.86	7.91
BP	4.41	4.82	5.26	5.65	6.04	8.18	7.96	8.10
Shell	N/A	N/A	N/A	6.54	N/A	10.8	8.43	9.00
Chevron	2.66	2.73	2.96	3.03	3.37	3.87	4.45	4.87
Conoco Phillips	0.47	0.47	0.68	1.10	1.23	1.64	2.27	2.66
Total	**13.66**	**14.27**	**15.11**	**22.83**	**17.75**	**34.0**	**30.97**	**32.54**

Source: Securities and Exchange Commission Filings, 10-K and 20-F.

Note: Royal Dutch Shell data is incomplete due to the administrative restructuring of the company. As a result, totals are not directly comparable.

Table 6. Shares Outstanding of Major Integrated Oil Companies (millions of shares)

	2000	2001	2002	2003	2004	2005	2006	2007
Exxon Mobil	6,936	6,809	6,700	6,568	6,401	6,133	5,729	5,382
BP	N/A	N/A	N/A	N/A	1,167	1,167	950	N/A
Shell	N/A	N/A	N/A	N/A	N/A	3,817	3,695	N/A
Chevron	713	1,149	1,137	1,137	2,107	2,232	2,164	2,090
Conoco Phillips	623	623	704	708	718	1,423	1,690	1,613

Source: Securities and Exchange Commission Filings, forms 10-K and 20-F and company annual reports.

Cash dividends paid reflect the product of dividends per share times the number of shares outstanding. For example, ExxonMobil' s 42% increase in dividends paid out between 2003 and 2007 represents a 42% increase in dividends per share only if the number of shares outstanding is constant. If the number of shares increases due to new issues, then the actual return is less. If the number of shares falls, the actual return to shareholders from dividend payments is greater. Thus, it is important to review the total shares outstanding, as shown in Table 6.

Table 7. Earnings-per-Share of Major Integrated Oil Companies (dollars per share)

	2000	2001	2002	2003	2004	2005	2006	2007
Exxon Mobil	5.04	2.21	1.68	3.23	3.89	5.71	6.62	7.28
BP	0.66	3.52	2.33	2.78	4.46	5.48	6.67	5.41
Shell	N/A	N/A	N/A	N/A	N/A	N/A	N/A	4.40
Chevron	7.97	3.68	1.07	6.96	6.28	3.68	7.97	8.77
Conoco Phillips	10.26	8.13	-0.57	6.91	11.60	9.55	9.66	7.22

Source: Oil Daily, Profit Profile Supplements, various yearly reports.

Share repurchase programs reduce the number of shares outstanding in the market, as the company buys back its own shares and keeps them in the company treasury. For the company, share repurchase creates a potential source of quickly available capital. Treasury shares may be resold in the market without further Securities and Exchange Commission registration and approval.

From the point of view of the investor, share repurchase programs should increase the capital value of the shares they hold, other things equal. This appreciation is due to the asset base of the company being divided among a smaller number of outstanding shares, in principle making each share worth more. Compared to dividends, there may also be tax advantages associated with this type of value appreciation.

ExxonMobil reduced its shares outstanding by 18% between 2003 and 2007. BP reduced its shares by 18.6% from 2004 to 2006, and Shell shares outstanding declined by 3.2% from 2005 to 2006. Although Chevron and ConocoPhillips data are more associated with share increases, these increases are likely at least partly associated with the mergers the companies were involved with in 2005 and 2006, respectively. In 2007, shares outstanding at both Chevron and ConocoPhillips declined.

Earnings-per-share is a popular indicator used by investors as a factor in determining the viability of a financial investment. During a period of share repurchasing, earnings-per-share reflects not only the earning power of the company, but the extent of repurchases. For example, earnings-per-share at ExxonMobil rose by 10% from 2006 to 2007, while earnings rose by 2.8% over the same period. As a result, the more than doubling of earnings-per-share experienced by ExxonMobil and BP, as well as the 4.5% and 26% increments at ConocoPhillips and Chevron, respectively, since the beginning of the current round of oil price increases that began in 2003, reflect not only market conditions but corporate strategy as well.

Debt and Capital Structure

Companies might use profits to alter their capital structure, defined as the balance of debt and equity financing. Corporate finance theory has long held that the choice of capital structure should have no effect on the value of a company in a perfect market.[5] In the real world of finance, however, where bankruptcy is a potential reality, many analysts look upon debt reduction as an important way to strengthen the balance sheet of a company, improving its financial health.

Although reducing debt might be a desirable corporate strategy, eliminating it entirely is not likely to be efficient. Debt financing, compared to equity financing, offers several attractive characteristics. Corporate interest payments are tax-deductible, while dividend payments are paid from after-tax income. This differential allows a greater net return from identical assets, one financed through debt, the other financed with equity. Additionally, issuing debt does not dilute the ownership base of the company for existing shareholders, nor does it dilute the capital value of an equity share.

Offsetting these advantages are several disadvantages. Interest payments are a fixed liability of the firm, unlike dividends, which are paid at the discretion of management. If interest payments are not paid on time and in full, legal consequences might ensue. Debt holders stand ahead of equity holders for repayment if the firm is forced into liquidation. Increasing proportions of debt on the balance sheet also make bankruptcy more possible, making the firm riskier.

Table 8 suggests that the major integrated oil companies seemingly have not followed the same strategy with respect to debt management. ExxonMobil has reduced its long-term debt from its peak in 2004, but long-term debt increased by 276% in 2007. BP, which had roughly steady levels of long-term debt in 2003 and 2004, saw increases in 2005 and 2006 and a 6.4% decline in 2007. Shell's limited available data show, again, a roughly steady approach until 2007, when long-term debt increased by 29%. Chevron reduced its long-term debt by 37% from 2005 to 2006, and by an additional 21% from 2006 to 2007, achieving its lowest long-term debt level in the data period. ConocoPhillips increased its long-term debt in 2006, again likely related to its purchase of Burlington Resources in March 2006, but longterm debt decreased by 12% in 2007.

Short-term debt is defined as maturing in one year or less. While short-term debt constitutes an immediate obligation on yearly earnings, short-term interest rates are usually lower than long-term rates, making this mode of financing cheaper in many cases.

Table 8. Long-Term Debt of Major Integrated Oil Companies (billions of dollars)

	2000	2001	2002	2003	2004	2005	2006	2007
Exxon Mobil	7.28	7.09	6.65	4.75	8.96	6.22	5.73	21.55
BP	14.77	12.3 2	11.92	12.86	12.90	13.92	16.72	15.65
Shell	N/A	N/A	N/A	9.10	N/A	9.91	9.58	12.36
Chevron	12.49	8.70	19.87	16.11	10.45	12.13	7.67	6.07
Conoco Phillips	4.13	8.26	18.91	16.34	14.37	10.75	23.09	20.29

Source: Securities and Exchange Commission Filings, forms 10-K and 20-F.

As shown in Table 9, the major integrated oil companies showed divergent policy with respect to short-term debt. ExxonMobil has reduced its debt every year since 2002 until 2007. BP saw expanding debt since 2003, but showed a 31% decline in short-term debt in 2007. Chevron showed low short-term debt levels from 2003 until 2005, but short-term debt turned upward in 2006, to decline by 46% in 2007. ConocoPhillips also accumulated short-term debt from 2004 to 2006. Longterm debt declined by 65% in 2007.

Table 9. Short-Term Debt of the Major Integrated Oil Companies (billions of dollars)

	2000	2001	2002	2003	2004	2005	2006	2007
Exxon Mobil	6.16	3.70	4.09	4.07	3.28	1.77	1.70	2.38
BP	6.41	9.09	10.08	9.45	10.18	18.01	22.34	15.39
Shell	N/A	N/A	N/A	11.27	N/A	34.44	6.01	5.74
Chevron	3.09	8.42	5.35	1.70	0.82	0.74	2.16	1.16
Conoco Phillips	0.26	1.12	0.84	1.44	0.62	1.75	4.04	1.40

Source: Securities and Exchange Commission forms 10-K and 20-F.

Table 10. Cash Holdings of the Major Integrated Oil Companies (billions of dollars)

	2000	2001	2002	2003	2004	2005	2006	2007
Exxon Mobil	7.08	6.55	7.23	10.63	18.53	28.67	32.80	34.50
BP	1.17	1.81	1.73	2.06	1.36	2.96	2.60	3.56
Shell	11.43	6.67	1.56	1.95	9.20	11.73	9.00	9.66
Chevron	1.89	2.21	2.96	4.27	9.29	10.04	11.40	8.09
Conoco Phillips	N/A	0.14	0.31	0.49	1.39	2.21	0.82	1.46
Total	**21.57**	**17.38**	**13.79**	**19.40**	**39.77**	**55.61**	**56.62**	**52.27**

Source: Securities and Exchange Commission Filings, forms 10-K and 20-F.

Cash

When revenues and profits accrue quickly, and perhaps unexpectedly, there may be little alternative to holding those returns as cash balances until a strategy for using them can be developed and put in motion.

As shown in Table 10, although the total cash held by the companies has increased by more than $32 billion since 2003, 66% of those holdings were in the hands of one company, ExxonMobil. At the other companies, the results are more mixed, with both increases and decreases observable.

While holding cash balances might be unavoidable in the short-run, a long-term strategy based on increasing cash holdings is likely to be questionable. While cash offers flexibility, it generally offers little or no return. The price for flexibility is the potential return lost by holding it. In the longer term, financial analysts generally agree that more productive uses should be identified for cash balances, or they should be returned to shareholders in the form of dividends if no feasible investment opportunities can be identified.

CONCLUSION

It is likely that the increases in the price of oil that began in late 2003 and persist today were unexpected by the major integrated companies. The onset of the war in Iraq and the rapid growth of oil demand in China, India, and even the United States, were not generally forecast. Hurricanes Katrina and Rita were associated with even greater degrees of uncertainty than the war and demand growth.

As a result, and especially in the context of 2002 being a relatively weak year for oil company financials, it is likely that no set plan existed for the use of the rapidly growing profits that began to accrue in 2004. In the several years that have passed since 2003, capital expenditures have begun to expand, investor returns have been enhanced, acquisitions have been made, and balance sheets have been strengthened, even though cash balance levels are still high and growing.

In time, as corporate plans more reflect a crude oil market characterized by higher prices, long-term assets might be accumulated, supplies of both crude oil and petroleum products might be enhanced, and consumers might see a slacker market where prices may moderate from current levels. Until that time, investors in the oil industry may continue to see high rates of dividend payout, stock repurchase plans, and the accumulation of short-term assets.

ENDNOTES

[1] Earnings include both domestic and international net incomes.

[2] Total SA, the French integrated oil company, is by some measures similar in size to the five companies analyzed in this report. However, Total's business interests in the U.S. market are limited, and for that reason Total is not included in this report.

[3] The five major integrated oil companies active in the U.S. market are ExxonMobil Corporation (ExxonMobil), BP p.l.c. (BP), Royal Dutch Shell p.l.c. (Shell), Chevron Corporation (Chevron), and ConocoPhillips.

[4] See CRS Report RL32530, *World Oil Demand and its Effect on Oil Prices*, by Robert Pirog, for a more complete analysis of the factors influencing world crude oil prices.

[5] Richard A. Brealey and Stewart C. Myers, *Principles of Corporate Finance,* McGraw-Hill, 4th ed., pp. 399-401.

INDEX

A

B

C

D

G

H

I

J

K

L

M

N

O

P

Q

R

S

T

U

V

W